Graphic Design in Educational Television

Graphic Design in

Beverley Clarke

Educational Television

Lund Humphries London

First edition 1974
Published by
Lund Humphries Publishers Limited
12 Bedford Square London WC1

Hard cover edition SBN 85331 359 8
Paperback edition SBN 85331 364 4

Designed by Herbert Spencer and Christine Charlton
Made and printed in Great Britain
by Lund Humphries, Bradford and London

Acknowledgements
The British Broadcasting Corporation; BBC/Open University Productions; Staff
of BBC/Open University; UNESCO; Centre for Educational Development Overseas;
University of Manchester/U.M.I.S.T. Audio Visual Service; Peter Montagnon;
Michael Philps; Robert A. Jones; Makoto Gotoda; John Aston;
Michael Graham-Smith; Alan Haigh; Brian Lee; Ivor Weir.

Illustrations by Haydon Young, Darrell Pockett.

Contents

Foreword

From the point of view of the television producer, television graphics are perhaps one of the most potent tools that he has at his disposal.

What then does he want – style, impact, wit? Yes, but above all, when he is making documentary or educational programmes, he wants clarity, economy, and purposiveness. The truism that genius is 1% inspiration and 99% perspiration holds good in this area, just as much as it does for the producer, director, or writer. Before you can deploy that essential 1% inspiration you have to do the spade-work – you have to know not only what the techniques are but how to use them. Above all you as a graphics designer must understand *why* the programme is being made, *what* it sets out to convey. Graphics work for television is not just the icing on the cake – it is part of the cake.

It is for these reasons that I support the approach that the author has made in this book. It does not tell you how to create award-winning opening titles – what it does do is to lay out, in a lucid and understandable way, the techniques that the graphics artist relies upon. The author is also a practitioner and I think that this shows through.

As I think the reader will see, the television graphics designer's work, given the very special canvas of the television screen, is vastly different in kind from that of the print designer. Above all, he has the inestimable advantage that his image can move, can grow, can change. He has a new element at his command – time. He can use rhythms other than the two-dimensional rhythms of the page.

For the designer who has not tried it, this is an exciting new world – one where he is required to be part film editor, part cameraman. To those about to take the plunge I wish the best of luck and don't forget what I said about the 99% perspiration.

Peter Montagnon
Head of Open University Productions

Chapter 1   **The Growth of ETV**

Graphic designers are trained and work in many different spheres of specialized activity – advertising, package design, book illustration, and so on. In most of these areas there is a substantial and evolving body of theory and practice which can serve as a guide for students and a stimulation to professional workers in the field. New ideas, new design styles, and new techniques are discussed in specialist magazines, and at professional conferences. The value of this continual dialogue is obvious to every designer who cares about the future of his craft, or even about its present standards.

Educational television is a comparatively recent development, with a rapidly growing demand for skilled personnel. ETV designers are beginning to realize that their specialism is sufficiently different from other kinds of design to benefit from internal debate as well as from contact with the wider profession. This book, which is intended primarily for other ETV workers, design students, and interested laymen, tries to provide a preliminary sketch of the philosophy and methods currently available to artists working in educational television. The final section of the book contains examples of work done in the field, together with brief accounts of the programmes represented. It is hoped that these brief 'case studies' will not only illustrate the sort of problems and solutions encountered in ETV, but will stimulate discussion about objectives and the means of reaching them.

Throughout this book emphasis will be laid on the importance of *function* as a key determinant of graphic style. Since a designer must first of all be aware of the purposes of his work, and since ETV establishments may serve any one of a range of different purposes, there follows a short (and necessarily over-simplified) survey of the uses to which television is put in education.

Table A: Approximate increase in population, 1930-70

Asia	868	million
Africa	280	million
Europe	115	million
North America	90	million
(World)	(1,500	million)

Over the last thirty years, several powerful factors have contributed to a world-wide expansion of *educational television* (ETV). It can be argued that all television is educative in some way, but the term *educational television* is usually taken to refer to a particular type of programme production. Frequently ETV is utilized and sometimes also administered by the formal institutions of education within a region; nearly always the programmes will attempt to communicate a structured body of information, ideas, or skills, and they are likely to be produced in series, often to supplement conventional courses of instruction.[1]

Since the educational uses of television are manifold, 'ETV' is an expression which describes not only a wide range of programme content but different *distribution systems* – from the small-scale closed circuit relay[2] to the national broadcast network. But the growth of ETV (at all levels of education) has been geographically widespread, even though the rate of expansion has proved uneven. How can we explain this growth?

The significance of the rapid increase in world population is widely recognized in relation to many urgent social problems. Table A suggests the dimensions of the increase between 1930 and 1970.

By the year 2000 the total world population is expected to have increased by something like 50% over 1970. In advanced countries the rate of growth is less dramatic, but its social implications are nonetheless important. With expanding numbers of pupils in schools and students in universities and colleges, there is a general feeling among politicians and administrators that additional student places should be provided at a lower cost per head than has been achieved in the past. Official attention has therefore been directed towards technological advances which offer the prospect of communicating information more cheaply.

Many countries have already experienced difficulty in training

[1] In the United States, the term *instructional television* (ITV) is used in this connexion, while ETV refers to certain categories of programme, such as serious drama, documentaries, classical music, etc, which are most commonly found on public service channels. In this book, the term ETV will be used consistently in the sense referred to in the text above.

[2] *Closed-circuit television* (CCTV) is a term used to designate those systems of distributing television pictures and sound where reception is restricted to TV receivers in direct contact (usually via cable) with the source. In contrast, *broadcast television* can be picked up by anyone with a suitable aerial and receiver.

Table B: Numbers of new book titles

Mathematics	1938	34
	1970	431
Sociology	1938	211
	1970	531

From *The Bookseller*

sufficient teachers to keep pace with demand, and recorded television programming has been seen as a method of releasing educators from repetitive large-group lecturing to concentrate on individual tuition.

Even if existing educational structures were able to accommodate increasing numbers of students without strain, governments would still face extreme difficulties unless willing to adopt new teaching techniques, because of continually rising levels of expectation. In advanced countries each generation has demanded both a more intensive and extensive educational provision for its children. This is, of course, in turn partly a reflection of the demands that increasingly complicated industrial societies make on their workers in terms of specialist skills. At all events, the standard of teaching itself must constantly be improved.

The developing countries often face graver problems. Lacking an extensive industrial base to generate greater amounts of wealth, the task of coping with explosive population growth and rising expectations must appear daunting. Not surprisingly, Asian and African educational administrators are looking more and more to radio and television for solutions to otherwise intractable difficulties.

If the numbers of people requiring an education (or a *better* education) are rising sharply throughout the world, there has also been an accelerating growth in the *amount of subject knowledge* available to teachers. Although the full significance of such figures requires a more sophisticated analysis than is possible here, it may be of interest to compare the numbers of new book titles published in Britain in two academic subject areas in a pre-war and a recent year (Table B).

The Director of UNESCO summarized all these trends when he wrote in 1967:
'Mankind is passing through a profound mutation caused by three

explosive factors, the increase in population, the speed at which certain knowledge becomes outdated and technical progress advances, and political emancipation. As a result, education must also undergo a radical mutation on a scale which can hardly, as yet, be fully appreciated. Many more people have to be educated for a continually increasing span of their lives so that they may absorb an ever-expanding and changing body of knowledge.

'It is impossible to conceive that these tasks can be undertaken without major changes in education. Fortunately, the need for such changes arises at a time when the media of communication, radio, television and film – and new methods and techniques of instruction, such as programme learning, have come on the scene.'

Before 1930 the word *television* was not to be found in any dictionary; by 1972 there were probably more than 250 million TV receiving sets throughout the world. Exploitation of the new medium's potentialities has been more rapid and thorough by the richest nations, and it was the USA that saw the initial development of television as a specifically educational instrument. Over twenty years of experience with other audio-visual techniques was applied to ETV. In 1945 the Federal Communications Commission, prompted by public concern, began planning for a widespread allocation of broadcast television channels. Seven years later 242 non-commercial channels were reserved for experiments with educational programming. Local education authorities and institutions throughout the country took advantage of the allocations to attempt to integrate TV instruction into their teaching systems. Not all these projects proved successful, but the early American experience – mistakes included – helped provide valuable guidelines, not only for subsequent developments in the USA but for educational planners in other countries.

The pattern of ETV usage within the United States partly reflects the decentralized character of the educational system. European countries with national television networks and more centralized control of education provide an interesting contrast. In Britain, for example, co-ordination of ETV services, though imperfect, is relatively far advanced; and students in any region of the country are able to draw on much the same type of ETV facility. This is certainly true of broadcast programmes. Both the British Broadcasting Corporation (BBC) and the Independent Television (ITV) commercial network have been producing programmes for schools and adult further education for many years, and these are available to any citizen with a television set anywhere in the British Isles. There is consultation between the two organizations about educational programming to ensure that duplication of courses is avoided.

Because of the almost blanket coverage offered by the broadcast networks, the Open University, the largest university in the world to award degrees mainly on the basis of teaching by correspondence, feels justified in spending a fifth of its annual budget on the production, in partnership with the BBC, of radio and television components for its courses. By 1974, 40,000 Open University students throughout Britain will be watching TV programmes as part of their studies.

British closed-circuit systems have displayed less standardization, but even here there are closer similarities than is usual in the USA. Many university systems, for example, are financed largely through the University Grants Committee, which advises recipients about the standards expected of their output. Moreover, where local education authorities have laid cable networks to carry programmes into the schools in their area (e.g. London, Plymouth, and Hull), there are similarities of usage which occur largely because of *national* policies for school education.

In Britain, as in other European countries, ETV has been used most successfully as a supplement to other forms of study. But Italy dramatically demonstrated the medium's potential for *direct* teaching with the foundation in 1958 of Telescuola, the television school. Soon the two-and-a-half hours of daily programming was reaching 50,000 students enrolled on vocational courses. On completion of a course, students sat final examinations – 70% of the candidates were successful. Telescuola later produced a series of programmes which – with the American *Sesame Street* – was to become probably the most famous ETV output to date: *It's Never Too Late*, which was produced in support of a national campaign to reduce adult illiteracy.

With its rapidly expanding economy, it was not surprising that Japan was able to invest heavily in the new technology. From 1961, ETV was employed to provide educational opportunities at upper secondary school level for children whose parents could not afford to send them to senior high schools. The Japan Broadcasting Corporation (NHK) began an exciting experiment in which a combination of correspondence material and broadcast programmes was used to provide opportunities for the educationally disadvantaged. This idea was, of course, later expanded and extended to higher education by the Open University in Great Britain.

Whereas industrialized nations have tended to employ ETV as a complement to existing educational structures, developing countries, faced with severe problems of financing additional schools and colleges, are often prepared to concede television a more ambitious role in teaching. The medium is seen as a way of extending educational provision at a far lower cost than would otherwise be required. In central west Africa, the state of Niger, with only sixty-six primary school teachers who had completed higher secondary education by 1965, was only able to accommodate 10% of the country's children in its primary schools. Classroom television, receiving programmes from a central production unit, was seen as the fastest way of boosting the pupil population, while at the same time releasing qualified teachers for secondary education. For the same sort of reason, Colombia in South America has organized a television network to provide instructional programmes for over 800 schools. Again, American Samoa has an extensive TV network providing televised lessons for several hundred schools. School buildings have been designed to accommodate this form of teaching.

Considered as a global phenomenon, then, ETV has been regarded as a device for communicating more information to more people. In recent years, however, a body of opinion has gained ground which holds that television can be used most effectively within an integrated multi-media teaching system. As a medium, television embodies characteristics which make it a suitable display for certain categories of information. The same is true of print, radio, film, teaching machines – and teachers! The conclusion is logical: each medium should be used to accomplish the educational tasks for which it is best fitted. (To date, there has unfortunately been little rigorous examination of the educational implications of media qualities.)

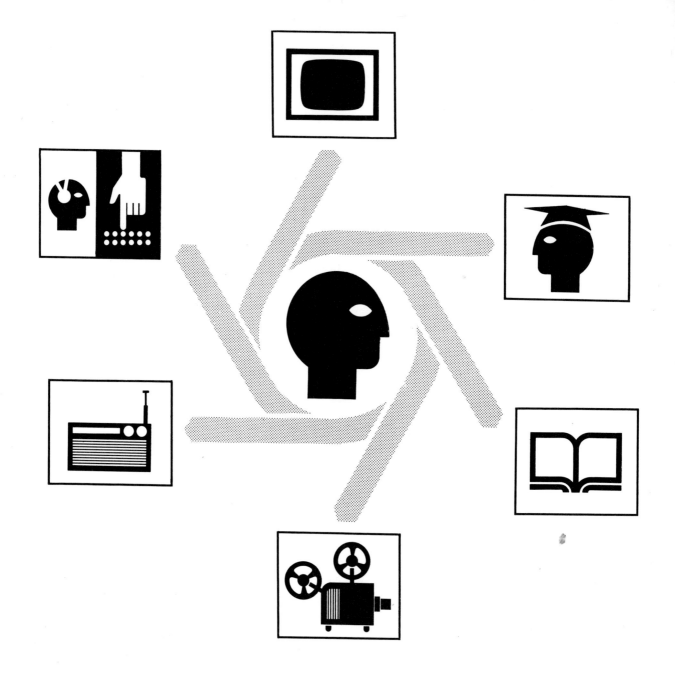

The importance of graphic design for ETV has never been doubted but, surprisingly, has rarely been discussed, at least in print. This is clearly a serious omission in a rapidly expanding profession which hopes to recruit from the best talent available. The subsequent chapters attempt to provide newcomers with an introduction to the field of ETV design, but it is also hoped that they may stimulate further debate among experienced practitioners.

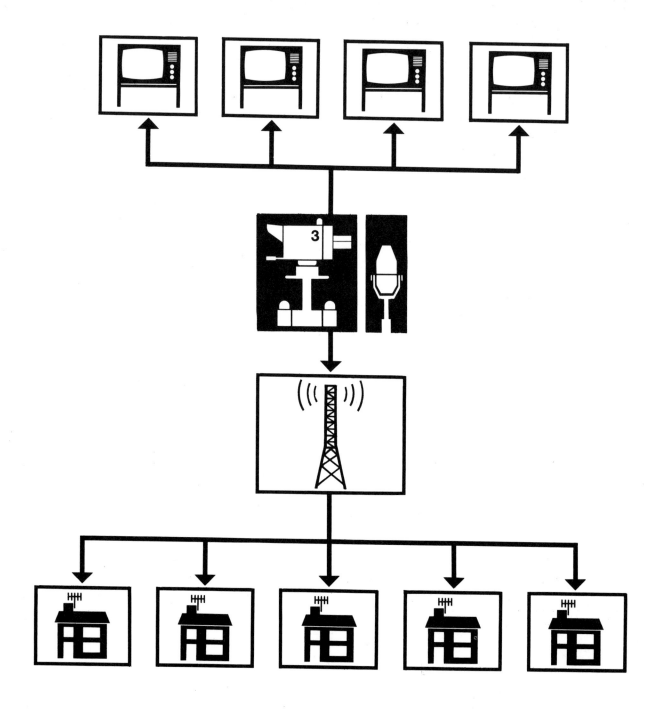

Chapter 2 ETV in action

The application of television to education has resulted in a wide range of different uses; from the broadcast transmission of elaborate and expensive programmes for many thousands of students, to the classroom use of TV cameras and monitors for magnification purposes. Television services performing different functions tend to be associated with particular *systems of distribution* by which video and audio signals are related to the viewer. These can be broadly categorized into *closed-circuit* and *open-circuit* systems. In the former case, signals are carried from television cameras and microphones to a limited number of TV monitors by way of coaxial cables or a microwave link. Audience size is limited by the necessity of establishing direct links to each TV receiver. In open-circuit distribution, signals are broadcast via transmitters, so that any TV owner with a suitable aerial and residing within transmitter ranges can obtain picture and sound images. If, as in most developed countries, transmitter networks have been constructed in order to service all but the most remote and sparsely populated areas, potential audiences may be numbered in millions.

Both open and closed-circuit systems can be employed at every level in a national educational structure, from primary school to university and beyond. The function of an ETV system and the resources available to it will help to determine the role of graphic design.

Closed-circuit ETV The simplest form of closed-circuit system comprises a single television camera linked by cable to a receiver. It is used in classrooms and lecture halls in order to show thirty or forty people small-scale objects, or experiments which otherwise only a few could see at any one time. A longer connecting cable allows the remote viewing of dangerous or inaccessible processes (maybe the

The addition of a videotape recorder to the basic system of camera and receiver enables unique or expensive programmes to be repeated for many different audiences.

interior of an atomic reactor) or of personal interactions, such as psychiatric sessions, which would be disturbed by observer presence. Obviously if more receivers are connected to the camera, larger numbers of students can be included, and several classes can, for instance, be taught by one teacher.

In this sort of system, graphic work, if used at all, will probably be restricted to titles and the labelling of objects, and may be prepared by the teacher concerned.

Another degree of sophistication is introduced into a system with the addition of further cameras and recording facilities. With three or four cameras at his disposal, a producer can select from (or 'cut between') several views of the same subject, or deploy the cameras so that they each cover a different subject. By means of a *vision mixing* mechanism the output of each camera will be chosen at some point during a programme as the picture which will be relayed to the viewers. Often, of course, a cameraman will be instructed to change his picture while his camera is not 'on air'.

A videotape recorder (VTR) records picture and sound signal onto magnetic tape, which may vary in width between $\frac{1}{4}$ in.–2 in. according to the picture quality demanded or the money available (high quality machines are very expensive). The VTR reproduces ('playback') the recorded material through linked TV receivers, and as with standard audio recorders, the tapes can be erased and reused.

This enlarged system allows the recording of unique or expensive events for future replay(s). Costly scientific experiments need only be performed once for teaching purposes because the recording can be played back to classes time after time. Important lectures by eminent visitors can be enjoyed over a number of years by a large number of different audiences. Perhaps most important of all, complete television programmes – which are very expensive items indeed, due largely to the man-hours involved in preparation, rehearsal and recording – can be justified on grounds of cost once VTR facilities have extended the potential audience.

The efficient production of programmes on a regular basis is usually facilitated by using a television *studio*. This will vary in size and complexity according to the functions of the institution which maintains it. Within a single school, for example, the television equipment may be stored in an ordinary room which has to double as a studio. Lighting and acoustic standards will be variable and camera movement drastically restricted. At the other extreme, national broadcasting organizations often possess purpose-built studio complexes costing several million pounds. But whatever form a studio takes, it will offer a convenient environment for graphic displays. *As far as designers are concerned, the most significant division between television systems occurs between those that possess studio facilities and those that do not.* Naturally there are wide disparities in quantity and quality of output between different types of organization, but a basic principle remains – the potential role of graphic material in education is so obvious to most teachers that they will not willingly forego its many advantages if there is an ETV environment suitable for its displays.

Closed-circuit television networks of different degrees of complexity are to be found in schools, colleges, and universities. Their functions vary according to institutional goals, but will probably include some of the following:

Overspill lecturing:
Instructing several classes, either simultaneously or (if VTR is used) over a period – the teacher exerting no greater effort than would be involved in a single lesson.

Observations of all kinds:
In teacher training, recording of classroom events will be a useful item for analysis and discussion; student psychologists will benefit from video-tapes of interviews and test situations when learning their basic techniques. The importance of observations for some types of education is so marked that some CCTV networks record their programmes from a *mobile unit* (i.e. a van or lorry housing the equivalent of a studio gallery and control room, and from which cameras can be extended on long cables) that can be driven to any location.

Visual aid usage:
As mentioned above, TV can be used to enlarge graphics, scientific experiments, or any other instructional material so that large classes can see them while remaining seated.

The production of specially-made educational programmes:
These will be comparatively expensive, involving several kinds of specialized skill in their preparation. Compared with, say, an observational recording, they will display a high degree of 'construction' and professional 'finish'.

Training in television:
As the medium has come to play an influential role in the leisure time of many advanced countries, there is a growing belief that education should encourage some critical response to the quality of national programming. Thus, some schools and colleges hope that their students will be less easily manipulated by the medium if they have gained insight into its workings by making programmes themselves. There is also a feeling that teachers who will utilize ETV programmes produced by others should have a similar training in the fundamentals of production, if only to alert them to the possibilities for distortion of the 'real world' inherent in the medium.

The staffing structure of a CCTV service performing some or all of these functions will depend on the *volume* and *type* of effort which the parent institution requires. Although there are exceptions, most set-ups will include a graphic designer if there are more than three full-time employees. The smaller the production unit, the more likely it is that a designer will be a jack-of-all-trades. Because of the need for speedily prepared graphics, for a flexible response to last-minute changes, and the cost of hiring freelance graphic artists and their equipment, the means of producing graphic materials will be on-site. The designers will be obliged to service a wide range of different types of television projects. This variety appeals to some professionals, and a high degree of expertise is often found in fairly small closed-circuit stations. Certainly a smaller and less specialized production team will tend to ensure that a designer will be more closely involved in the production process from the earliest stages.

When designers become more specialized – as in the large national broadcasting networks – specialization is usually determined by technical expertise, and the artist is called upon only when his particular skill is required. This means that it is, for example, more difficult for an artist to become accomplished in the field of educational graphics than in that of animated film or typography. In fact the larger organizations rarely encourage their designers to specialize by *programme content*, and a natural consequence is the educational producer's reluctance to consult his designer until the important decisions have already been taken. This will rarely happen even in the largest closed-circuit ETV units (and some of these, like that run by the Inner London Education Authority, are large enough to link a sizeable fraction of the state-run educational institutions in a large city area) because designers are experts in *educational* graphics as well as educational *graphics*.

Broadcast ETV   In most cases broadcasting networks, whether publicly or commercially sponsored, are available to large publics – sometimes to the population of an entire nation. Education is rarely the primary function of such a network, though the instructional opportunities inherent in it are considerable. Programmes can be transmitted to a large potential audience at a very low cost per head. The financial constraints on programme makers will probably be less severe.

Educational programme production will usually constitute a separate unit within a broadcasting service. Its producers and directors will be specialists in education, but for its other technical and creative resources it must draw upon service departments which provide skilled manpower (on a short-term basis) for every type of programme the network produces. Designers working within an organization of this sort may have been recruited especially for educational programmes, though this is unlikely. Alternatively, a design department may encourage its personnel to specialize in certain areas of programme making, and by mutual consent a designer might find himself working almost continuously with one production department. There is some conflict of view amongst designers about the desirability of specializing in educational graphics. On the one hand it is generally recognized that ETV design requires many unique skills, but on the other it is feared that creativity will somehow be blunted if a designer remains too long 'out of touch' with mainstream graphics.

A broadcast network designer finds a high level of sophistication in the technical resources available for the production and presentation of graphic materials. Specialized facilities and services are not necessarily on-site, and having visualized his material, a designer may contract out the physical preparation of artwork, film animation, studio displays, etc.

The main constraint (apart from organizationally-defined work roles) affecting the designers will be the limited time allocated for completing his work on any programme. Failure to meet 'recording deadlines' can result in the loss of large sums of money if studio time has to be cancelled at short notice (and thus wasted).

A large national network currently represents the most extensive system of ETV. In the next few years, however, the evolution of *international* services, based on satellite transmission, may be expected.

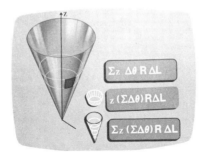
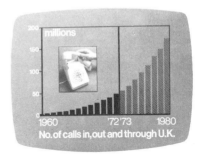

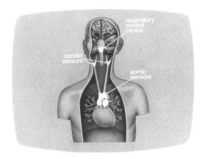

Although the word *design* is used in many different fields (e.g. in graphics, engineering, the stage, fashion, architecture, etc.) one underlying principle seems to hold true: design encompasses those qualities of an article which contribute to its effectiveness in manufacture or use. Any person whose work involves the visualization of these qualities before the object itself is produced, may be called a *designer*. The definition of these two words is sometimes confused by references to a purely 'aesthetic' dimension of design. In fact, if this term is to mean anything, it can only refer to the feelings of satisfaction we experience when using an object that performs its function efficiently.

Graphic design is concerned with the communication of visual information, comprising either typographic (lettering) or pictorial elements. A wide variety of media is used to convey graphic information: books, magazines, posters, packaging, films, television, etc. Within each medium there may be many different *kinds* of information requiring graphic expression, but the designer's task will always be to present it to the best advantage. In order to achieve this goal, the designer must be fully acquainted with the physical characteristics and the creative potentialities of his medium.

In educational television, graphic design is only one of several techniques used in programmes to impart facts or ideas. Graphic material may take a variety of forms, from abstract imagery, such as graphs, tables, mathematical formulae, and histograms, to more representational diagrams such as those illustrating scientific apparatus or human anatomy. In every case the designer will attempt to evolve a style which will enable the viewer to comprehend and assimilate the information as rapidly as possible.

Because the work of the graphic designer in ETV often performs a more crucial function than that of his colleagues in entertainment programming, he is likely to be more heavily involved – from the early stages – in the planning of a production. He will, therefore, have greater opportunities for making an important contribution to the central purpose of a programme, even though he may be required to do so on a more restricted budget. His success in this will depend partly on his grasp of the relevant technical characteristics of television.

Aspect ratio This term refers to the horizontal and vertical dimensions of the television screen. The width to height ratio is always 4:3 on modern receivers. Screen sizes are measured along a diagonal line from corner to corner.

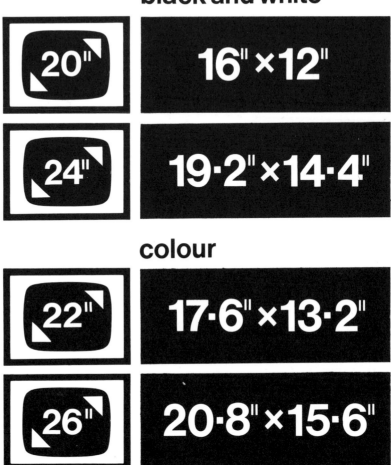

black and white

| 20" | 16" × 12" |
| 24" | 19·2" × 14·4" |

colour

| 22" | 17·6" × 13·2" |
| 26" | 20·8" × 15·6" |

All design artwork must always be prepared in the 4:3 ratio, and in the case of photography, it is advisable when using cameras to remember that the compositions will finally appear in this format.

Line structure   A television picture comprises a series of fine, horizontal lines. According to the standard set by a particular country, there may be 405, 525, 625, or 819 lines per picture. Other things being equal, the

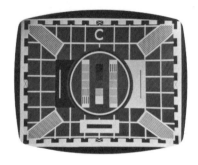

more lines, the better the picture resolution (i.e. its ability to reproduce detail). Even the best TV systems are incapable of reproducing delicately drawn lines or shading. Type faces with slender elements may suffer distortion or complete illegibility; closely spaced lines or letters will tend to merge. It is therefore essential that the designer be acquainted with the constraints imposed by a particular line standard. A full appreciation of the practical implications can only be acquired with experience, but the beginner should try to obtain access to a series of different TV test cards and compare their appearance on and off screen. It will also be useful to experiment in the same way with photographs, for again some textures and details will certainly be lost.

Cut off Compared with studio equipment, most television receivers are manufactured to fairly low technical specifications. Their capacity for displaying the *total* picture as it leaves the studio will vary considerably. Television engineers and producers are usually disinclined to take chances on the complete accuracy of picture reception, and allow a 'cut off' area to safeguard against picture loss in transmission. This may mean that as much as 20% of the total picture area is treated as 'out of bounds' for information that is regarded as important. Sometimes a 'safe area' is actually marked on studio monitors so that a director can be sure that essential material will reach his audience. It is always important for a designer to produce artwork which allows for the appropriate cut off, but in ETV, where the graphic information often constitutes a vital part of the programme, special precautions must be taken. A safety margin should be left round the edges of artwork and photographs, so that, for example, on 12×9 in. artwork the actual design area may only be $8\frac{1}{2} \times 7\frac{1}{4}$ in.

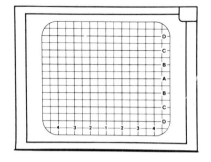

Screen size Whatever the future may hold, television pictures are currently viewed on a small screen. In fact, the average screen size viewed at an average distance is only equivalent to a 12×16 cm sheet of paper held at arm's length. Since the screen fills only a small area of our total field of vision, the legibility of type or other graphic material is determined partly by the distance from which the picture is viewed. This will vary from perhaps 8 ft in a domestic setting to viewing distances of over 30 ft in classrooms and lecture theatres (see page 76). Some teaching institutions possess electronic devices to project television images onto a film-sized screen, though these are comparatively rare and generally unpopular with designers, since the enlarged line structure highlights the coarseness of the television picture. At all events, the graphic designer is well advised to discover all he can about the screen sizes and the conditions under which his work will be viewed. The designer who creates merely with the appearance of his artwork in mind can never be an adequate *television* professional.

Tone scales The range of tones (between black and white) which television cameras and receivers are capable of reproducing, is known as the 'grey scale'. A high-quality television system can distinguish as many as twenty different greys, but this range will be reduced considerably upon transmission and reception of pictures. Artwork for ETV often incorporates as few as five tones because the receiver's performance will be variable and the graphic material essential to the programme.

The use of colour alleviates many of the problems of black-and-white television and increases the potential differentiation between different areas on the screen. However, in most countries the designer must remember that his work will still be seen on many monochrome receivers. His choice of colours must, therefore, relate directly to distinguishable tones on the grey scale. A dark blue will, for example, read as dark grey in black and white, while a pale blue will appear as light grey. Even when a designer knows that all receivers are colourized, he will do well to investigate the characteristic limitations of his system if he intends to use delicate gradations in his artwork. It cannot be over-emphasized that an ETV designer is principally concerned with communicating meaning. He

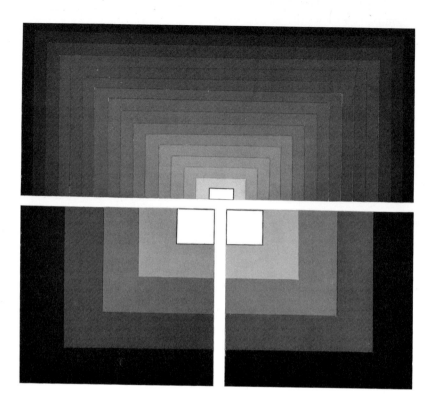

Although a wide range of greys can be distinguished by a high-quality TV system, artwork for ETV often incorporates only a few tones to allow for limitations in the quality of reception.

has failed if his work is illegible through maladroit layout or tonal separation.

The next part of this chapter will deal with apparatus used to display graphic materials for the television camera.

Studio apparatus   There is some debate among educational broadcasters as to whether or not the television teacher is more effective if seen in personal association with his illustrative material. Some producers believe that a performer should be able to point to diagrams and manipulate pictorial elements in a diagram; others insist that a presenter be separated from his visuals in order that student attention can be focused without distraction. In either case, studio

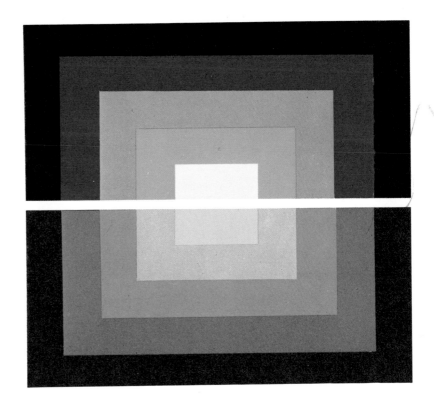

Artwork for colour programmes must still relate to distinguishable tones on the grey scale to allow for reception by monochrome receivers.

A caption refers to any type of graphic artwork or photograph which is mounted on a piece of card. Captions will often be 12 × 9 in. but will rarely exceed 30 × 20 in. as the heat from studio lighting will buckle card of standard thickness – and heavy card will reduce the number of captions which can be placed on a stand.

apparatus is available for a designer to display his work. Where the visuals are to be isolated from the presenter – that is, as 'cut-aways' – the most common device will be some form of *caption stand*.[1] Stands are constructed to hold captions at a height and angle convenient for the TV camera and studio lighting. They should be securely based to withstand vibrations in their vicinity, yet easy to adjust. Captions must be held parallel to the studio floor lest horizontal lines on the artwork be skewed.

Larger stands such as *strap easels* are used to display studio animations (see page 35) or captions which need to be larger than 12 × 9 in. so that the camera can explore their detail by panning or zooming. *Roller caption stands*, which can present a continuously

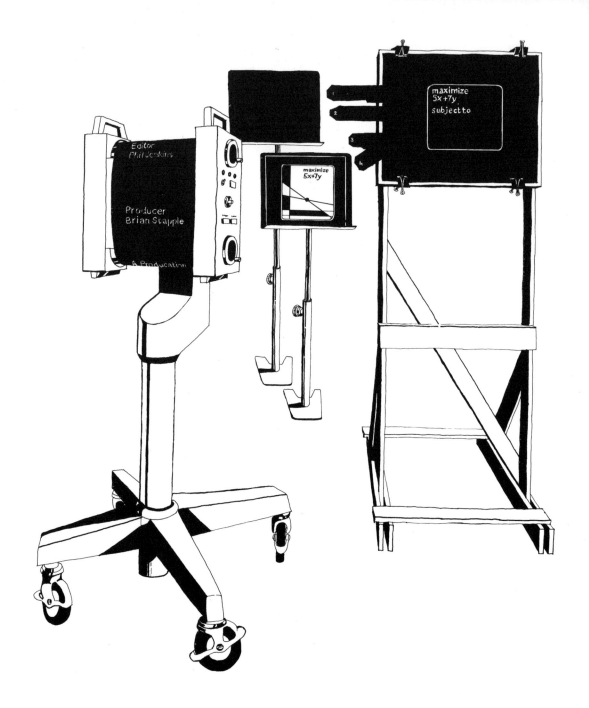

moving graphic, are most commonly used for programme end credits. The stand comprises two motorized rollers between which passes a roll of paper bearing drawn or printed lettering. The camera remains stationary, shooting the area between rollers.

Apparatus intended to display large-scale graphic materials should be integrated into the general scenic design of the studio, particularly if the programme presenter is to appear in the same shot. Display boards will usually be somewhere between 4 × 3 ft and 8 × 6 ft. in size and can be constructed to revolve or to incorporate sliding panels. A wide variety of materials are used for the surfaces of display boards: painted plywood, tinplate to which magnetized cut-outs will adhere, or sheet plastic serving as a writing area.

Lighting engineers will take particular care to achieve uniform illumination for these large displays and the designer can help by selecting non-specular materials.

The preparation of artwork on this scale (see page 50) will be comparatively expensive and time-consuming. To some extent, photographic processes can be employed to avoid additional costs. Artwork of the usual size can be prepared for photography, and in the case of monochrome TV systems, a large *photo blow-up* can be printed for studio display. The cost and quality of colour blow-ups will prevent their use, but *projected photographic images* can be utilized for coloured graphics of large area. A *back projection* (BP) unit consists of a high-powered projector which throws an image of a photographic transparency on a translucent screen. The television camera views the screen from the opposite side, and with careful studio lighting a presenter is able to appear in the same area.

Back projection offers important advantages for the small television studio because a series of large-scale graphics can be displayed without the necessity of moving cameras or presenter. Many designers, however, are reluctant to make use of projected images because of an inevitable loss in resolution and the difficulty of achieving an even intensity of light on the BP screen. In sophisticated colour studios, back projection has been displaced by an electronic process (see page 38) which accomplishes the same effects but with superior quality.

Electronic apparatus

Floor space in a television studio is often at a premium. Apart from television cameras, which require substantial areas of clear space for rapid manoeuvrability, there may be sound equipment, floor lighting, studio monitors, models, scientific apparatus, a studio set, graphic displays, presenters, and a team of technicians to operate the equipment. Machinery which can exhibit artwork without occupying floor space is, therefore, often a welcomed resource. Sometimes, too, these devices offer a unique facility or a quality of image unobtainable in any other way.

Slide scanners are sometimes regarded as an alternative method of displaying information which might otherwise be prepared in caption form. A slide scanner usually comprises two 35 mm slide projectors and a fairly simple television camera. An optical system enables the camera to view illuminated slides from each projector in turn, and since the resulting picture signal is passed through the studio vision-mixing panel, cuts and dissolves between slides can be achieved. There are, however, three principal drawbacks. The first holds true of all photographic materials, particularly slide transparencies: they are difficult to modify at short notice. More important from the director's point of view is his inability to explore the image's detail by zooming or panning his cameras. Finally, no designer should rely on the extensive use of slides unless he can call upon the services of a highly-trained photographic department. The density of the image must be carefully controlled in processing, and slides should be mounted in a dust-free environment (the picture may be magnified up to twenty times) with equipment that guards against misalignment.

Telecine equipment, which converts projected cine-film images into television pictures, is regarded as essential by most ETV organizations. For the graphic designer, it opens up the possibility of working with film; a valuable resource, particularly for animated sequences (see page 55).

Other electronic devices can be used to produce television pictures which are assembled from different sources – it may be two television cameras, film from telecine plus superimposition from slide, or many other combinations. It is increasingly recognized as an essential part of a television designer's expertise to be able to create 'special effects' sequences using this machinery.

Superimposition is used to mix an image from one source with a second image from a different origin. It is most commonly used for 'labelling', for example, identifying a studio speaker by superimposing his name. Because superimposed graphic material is usually an additional element in an existing scene, the latter is

Concerto Grosso
Fewer passages for the *concertino* group alone; musical interest shared more equally throughout the orchestra

referred to as the 'background'. Lettering or drawing for superimposition must be prepared on a black background, and in monochrome systems should be in white.

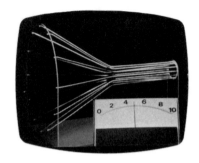

Inlay is a method by which part of one television picture is electronically 'cut out' and then filled in by visual material from another source. The outline of the cut out area is determined by the particular inlay mask used. In ETV frequent use of the inlay technique is made to achieve 'split-screen' effects. Here the screen is divided so that information from different sources may be compared: perhaps a working scientific model alongside a graphical representation of the same object, or two animated graphs demonstrating the effect of one economic variable on another. A director may wish to use split screen when he needs to present more than one item of visual information without intercutting between cameras. Graphics for split-screen effects have to be meticulously designed to fit previously agreed framings, lest the information either overlaps or disappears off-screen.

Overlay works on a similar electronic principle to inlay, but the foreground subject from one television camera can be inserted into a background picture provided by a second source (camera, slide, or telecine). For example, lettering will be 'overlaid' into a picture when the background is too light for a white superimposition to read. Here the lettering itself is the mask which cuts into the picture, and can then be filled in with a plain black or grey tone from another source. However, typefaces tend to suffer distortion through this process and this restricts a designer's choice. In colour television, the technique can be extended to use a presenter in the foreground, and is termed colour separation overlay (CSO) or chroma-key. The foreground subject is placed in front of a highly saturated blue background, this colour activating the keying switch in the overlay device, which results in the blue area of the foreground picture being replaced by pictures from a second source. Since CSO draws on 'direct sources' for its backgrounds, it produces higher quality

Example of colour separation overlay.

composite scenes than studio back projection. Perhaps its greatest potential usefulness for ETV lies in small studios, where a presenter in a single location can be shown in association with an unlimited series of high definition pictures, both still and animated.

Reverse scanning is an electronic mode which laterally inverts (or produces a mirror image of) a television picture. This facility is particularly useful when, for practical reasons, a camera is compelled to view a scene through a mirror.

Reverse phasing electronically inverts the tonal scale of a monochrome television picture so that white becomes black, and

Reverse scanning facilitates lateral inversion (in other words, a mirror image) of a subject where required.

Opposite page: videotape can be edited electronically to achieve video animation.

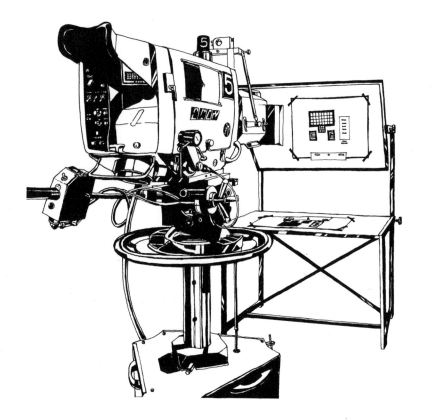

vice versa. This can sometimes assist the graphic designer when he is using materials which are easier to handle in a tone directly inverse to that desired in the programme. For obvious reasons, the facility cannot be invoked if a presenter is to appear in shot.

Video-animation is a more recent development, but has great significance for the designer. A television camera is used in a similar way to a film animation camera, recording a scene into which new pictorial elements are successively introduced, producing the impression of movement. One of the advantages of this resource over film animation is the facility of instant playback via a VTR machine (see page 21). The designer can see the animation

developing at each stage during recording and modifications can be made by erasing an unsatisfactory shot and starting again. Because video-animations are achieved by a direct camera source, their quality picture is noticeably better than artwork recorded through a photographic process such as cine-film or slides.

Video-animation, however, depends upon the capacity to edit video-tape electronically (and by no means all VTRs possess this facility) and the quality of the finished animation is partly dependent on a highly skilled engineering staff.

It cannot be over-emphasized that these devices can only be exploited to their full potential by the designer who acquires some technical knowledge of their workings, and who regularly takes the trouble to observe them in operation.

Chapter 4   **Design Preparation**

Typography

Educational material, by definition, has a high informational content. Apart from pictorial graphics, much of the *explanation* of an ETV programme will be achieved through spoken language. There will be many occasions, however, when written language, in the form of typography, can be used to advantage. Among other things, type can be used for:

opening and closing programme titles;

the identification of programme participants;

the emphasis of unfamiliar words, key ideas, or definitions in the verbal commentary;

labelling diagrams and models;

the presentation of mathematical or other notations (e.g. algebra).

The first consideration in using type must always be its legibility on the television screen, and this often presents major design problems, especially when graphics are required to carry a heavy informational load. It is probably not overstating the case to say that the selection of suitable type sizes, and type faces, and their layout is one of the most important skills demanded of an ETV designer.

The print medium has at its disposal several thousand different typefaces, these having evolved over hundreds of years. Unfortunately, a large proportion of these styles cannot be used effectively on television. Until recently designers have had to be content with a limited range of faces which are legible on the small screen, but there have been attempts to develop special typefaces for televisual use, and one of the most recent developments bypasses the use of an actual type image as artwork, and makes use of the resources afforded by computerized typeface storage facilities.

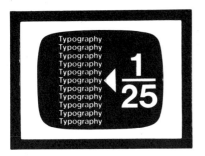

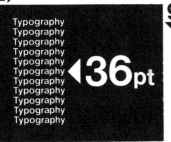

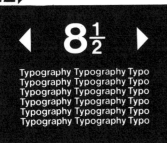

The principal factor governing typographic *layout* on ETV is the minimum size of type needed to achieve legibility in specific viewing conditions. For domestic viewing, the minimum typographic character height will be no less than 1/25 of the total picture height.[1] This means that on 12×9 in. artwork, the minimum size of type which can be used is 36 point. The minimum type size in turn helps to govern the *amount* of typographical information that can be accommodated on the screen. Another factor here is the safety margins which are left to allow for 'cut off'. So the length of line on 12×9 in. artwork will be no more than $8\frac{1}{2}$ in. long.

Having established the minimum type size and maximum line length does not leave the designer free to squeeze as many words as possible into a line by closely packing the letters. Such a line would merely appear as an undifferentiated mass, due to the line structure of the TV image (see page 29). Cramming lines together in the hope of increasing the amount of information would produce the same result.

Apart from these technical constraints imposed on typographic layout, designers in ETV often favour other general principles, such as the restricted use of words in capital letters (lower-case type being generally more legible), and the attempt to ensure clarity in layout design by limiting the combinations of *different* type styles and/or sizes in the same picture frame.

Typographic design is also affected by another less tangible factor: the purposes for which typography is being used. Consider the following:

(a) Programme titles need to communicate information quickly, if possible in a way which arouses interest, yet at the same time must not violate the approach or 'mood' of the programme.
(b) Type for superimposition, inlay or overlay needs to be bold enough to stand out from a tonally varied background, but not so

[1] Type sizes are measured in *points*. A point is approximately equal to 1/72 inch; so 72 points will be an inch high.

Not all typefaces developed for conventional printing are suitable to be taken over for use on the TV screen. Constraints on legibility involving type size, line lengths, viewing conditions, and line structure of the TV image must be taken into account.

bold as to distract students from objects of central concern.

(c) On diagrams, typographic style will be determined by its overall function within the greater whole. Is it to provide the diagram with a title; is it to label important elements; or is it in itself an element?

(d) Lettering on models and large-scale studio graphics will partly depend on the kind of shots the director intends to select. In other words, it is necessary to discover beforehand exactly which framings will be used, because only then can appropriate type sizes and arrangements be planned. If, for example, a director wishes to 'explore' a model by using a series of shots taken from different angles and at various distances, the designer must devise a labelling scheme in which type performs its function on each individual shot, but also avoids interference with other shots.

(e) In the case of mathematical notations, type style should, within limits imposed by the medium, follow the known conventions of the discipline or the usages adopted elsewhere in the students' learning materials.

The limitation on the amount of information which can be simultaneously displayed on a television screen poses problems familiar to all educators working in ETV. Information must be broken down into manageable units and a strictly sequential presentation of graphic material is often inevitable. The teacher will often feel uneasy that his students cannot refer back to information they have seen previously, and certainly more is being expected of a student's memory than when he reads a book, which offers instant access to all the information it contains. The designer can help his director to diminish these problems in several ways. He can, for example, help to ensure that information is segmented so that the student's attention is engaged and directed in such a way that he fully comprehends and memorizes each component 'frame' in a series. Design can also be employed to help integrate a series, because viewers will more readily identify different visual components as part of the same exposition if they can recognize elements presented in a consistent and distinctive style.

The technical side of type *production* involves the designer in decision-making of an easier, but no less important, kind. The main factors here are usually *cost* and *volume of output*.

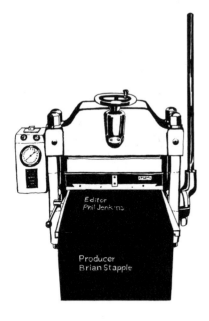

For producing large quantities of work which is of a standardized nature (i.e. utilizing a limited range of typefaces and layouts) designers use a *hot-press machine*. In this equipment, heated lead type is pressed face down onto a sheet of thin plastic foil which covers the caption card. Under pressure, ink is released from the foil and leaves an impression of the type layout. Although simple in conception, this machinery does require skill in operation, for unless the temperature and pressure are gauged exactly, there may be excessive ink loss or deep card indentation. An experienced operator, however, will be able to produce good results even on materials like cel acetate – which may prove useful for film animation.

Smaller quantities of work or lettering which needs to be placed on more difficult surfaces is usually accomplished with some form of *letterpress transfer* technique. The letter-face transfers are commonly purchased on sheets of thin plastic. Individual letters are transferred onto the artwork, according to a prepared layout, merely by rubbing the reverse side of the sheet. Wide ranges of typefaces and sizes are available on these transfer sheets, providing enough choice for most ETV needs. However, as we have seen, if each letter were to be transferred individually a large throughput of similar typographic work would involve a time-consuming, and thus uneconomic, use of a designer's effort. Also, very large or very small letter sizes are often difficult to manipulate using transfers.

Typographic elements for large studio displays can be produced quickly by using self-adhesive letters. These are usually made of plastic, and are relatively expensive to buy, though with careful handling they can be reused.

Where special typographic effects are needed (such as the distortion of lettering), *photo-typesetting* can be used. Here the letters are stored on film and printed by photographic techniques in such a way as to produce the required forms. The method is usually considered too expensive for more than occasional use by ETV.

Diagrams   In the widest sense, the functions of diagrammatic design in educational television are:

to compensate for the limitations of the television image;

to present information in a style which can be easily assimilated and understood;

to relate graphic elements to other programme components (e.g. verbal commentary, actuality visuals, etc.) to help produce an integrated whole.

In order to achieve these objectives, the designer must be fully conversant both with the ideas to be communicated and with the

A representational diagram, based on a pictorial convention which most people would recognise.

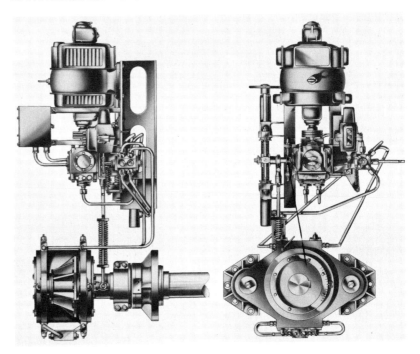

teaching aims of the programme. It is desirable that he be consulted from the earliest stages of programme preparation and be given the opportunity of offering suggestions for pictorial interpretations.

Elements on ETV diagrams can be divided into two broad categories The first we may term *representational* because the intention of the visual is to demonstrate something about the structure or functioning of a real entity. Obviously diagrams of this sort always rely on simplification and abstraction, though the degree will vary, as will the availability of existing pictorial conventions. The illustration on page 47, for example, shows representational conventions which most people would recognize because the shapes bear some resemblance to the objects they represent. The diagram above, on the other hand, illustrates certain relationships within a film production unit and uses symbols which bear a more tenuous connexion to the originals. The more we move in this direction, representing increasingly abstract concepts with graphic

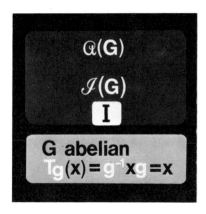

elements which look less and less like real objects, the closer we come to the second category, which we will call *notational*. Notations are series of arbitrary symbols used to stand for certain completely abstract ideas. The concept of number, for example, can be represented by one notation which begins 1, 2, 3, 4, 5 . . . Notational diagrams will include graphs, statistical tables, mathematical formulae, etc.

The over-riding consideration in the design of notational diagrams is likely to be legibility: since the 'meaning' of the visual is embodied in the notation, a designer's creative role is limited. Representational diagrams offer more scope for the artist, particularly if he has become an acknowledged member of the production team.

However, diagrams having been conceived, their physical preparation should be accomplished in the least time-consuming and expensive ways consistent with a result of the appropriate quality.

Artwork for captions is usually prepared straight on to card; intermediate stages are unnecessary because television's comparatively coarse picture does not demand a high-quality finish. The most common type of caption card has a fairly fragile surface and excessive use of rubber pencil erasers should be avoided. If the card surface is damaged in this way, the marks will appear as tonal blemishes when the caption is seen through a studio camera. Yet if pencil guidelines are left on the caption, studio lighting will probably reflect them. An answer to this problem can be found in the use of paper strips as guides for straight lines.

Where large areas of flat, unshaded colour are required on a diagram, cut-out pieces of art paper may be used in preference to paint, in order to save time and reduce the effect of errors. However, it is inadvisable to build up too many layers of artwork because the caption may become difficult to light evenly, and also the top layers will be susceptible to damage if sandwiched between other cards on a stand.

Some of the materials used by the TV
designer in the preparation of artwork
for large displays; notice the diagram
to be scaled up to 4 ft square.

Artwork for large displays is also best prepared straight on to the
final surface, though the layout should be planned at a convenient
size and then scaled up to the larger dimensions. With the
large surface areas involved, bold techniques for the creation of
artwork need to be adopted. Pictorial elements which will adhere to
the surface instead of being painted are to be preferred. Cut-out
paper shapes instead of painted areas, self-adhesive chart tape
instead of drawn lines, self-adhesive letters and symbols – all are
materials which will help the designer reduce his preparation time.

As with other kinds of artwork, time and money can easily be wasted
by providing a 'finish' which is beyond the capability of the
television system to reproduce. When an experienced designer has

ascertained (early in the planning process) how closely the cameras are to explore each area of a studio display, he will know with what standard of finish the artwork must comply. For example, a large studio graphic seen at a distance with a presenter will not require the same fine detail as sections of artwork which are to be isolated in close-up.

Artwork for projected display will need a high quality finish both because any blemishes will be magnified upon projection and because the picture quality loss sustained with most studio projection devices must be compensated for as far as possible. Artwork should be kept flat so that lighting for photography presents no problems, and some sort of framing guide should be provided for the photographer.

To save processing time and conserve quality, artwork for back-projections or slide scanners is sometimes prepared in 'negative' form. The photographic negative is then used for projection, rendering the original artwork as 'positive'. This procedure is only possible with black-and-white television systems. Moreover, the preparation of artwork can be confusing if complicated compositions using different tones are involved. The most effective safeguard against errors of tonal rendition lies in the use of a reference scale with labelled tones that will indicate precisely which 'negative' values correspond with which 'positive' tones.

Colour slides need never be prepared in this fashion because there is available a wide range of colour reversal film stocks.

Artwork for use with electronic special effects must necessarily be of a more experimental nature because much of the equipment concerned has been introduced only recently, and its full potential has yet to be exploited by designers, especially in educational television. Nonetheless, one vital procedural principle can be stated. While in monochrome systems a designer knows that material for superimposition or inlay will usually be white on black or (less frequently) black on white, colour pictures add an additional element of complication. The colour composition of the background shot must be determined before the overlay insertion can be designed, if an unpleasing combination of colours is to be avoided.

Animations: three types

Techniques of animation are employed in ETV when it is necessary to demonstrate *processes* by showing pictorial elements in dynamic inter-relation. Diagrammatic movement can be used to highlight changes in size, speed, or density; to illustrate direction or flow patterns; to delineate sequential patterns of a spatial or temporal kind; or merely to direct the viewer's attention to different parts of a diagram.

The three types of animation in most common use by ETV producers are magnetic diagrams, animated captions, and filmed animation. Each method has its characteristic advantages and drawbacks.

Magnetic diagrams

Where animation is required to show the accumulation, subtraction or substitution of pictorial elements, and especially when it is felt that the programme presenter can best contribute by actually manipulating the diagrams he is explaining, this technique can be used. A 'background' drawing is executed on a surface, such as tinplate, to which strips of magnetized rubber will adhere. Movable pieces are usually drawn on card which has been pre-cut to the correct shape and backed with the rubber. During the programme a presenter will move or replace cut-outs as appropriate to his verbal commentary.

From the designer's viewpoint this sort of device has only limited application because the size or shape of cut-outs cannot be varied in shot and, more importantly, even the simple movement which is possible may be obscured by the presenter's hands. Producers will also be cautious in their use of magnetized diagrams, as lengthy studio rehearsal periods are often required.

However, of all animation devices, this demands less specialized skill in design. Two simple principles are worth mentioning: On monochrome systems, avoid tones at each extreme of the scale, otherwise studio engineers may experience difficulty in coping with shots including human 'flesh tones'. If the background is light in tone it will probably be susceptible to marking during rehearsal unless sellotape is stuck over the magnetized rubber in contact with the tinplate (this will not seriously affect its magnetic properties).

Animated captions

These are constructed from several layers of card, and are operated manually, according to simple mechanical rules, outside the camera field area. The basic principle involves the uncovering or concealment of graphical information on the various layers. When pulled away, successive series of movable paper panels will reveal pictorial elements through cut-out 'windows' in the higher layers.

Animated captions can only be really effective in monochrome because their workings could be seen by the viewers were it not for the studio engineer's ability to compress the grey scale to two tones, black and white. To the naked eye an animated caption appears as a mass of cardboard 'windows' and masks, with deep shadows clearly visible between layers. When the picture has been electronically adjusted, the background appears as an even black, while the white elements appear – apparently from nothing – as the operators manipulate the sliding panels. The range of possible movements is determined by the mechanical structure of the caption, and considerable ingenuity is sometimes exercised in achieving fairly complicated movements. However, the more sophisticated the structure, the more delicate it becomes, and the more difficult to operate. Each caption can therefore only incorporate a limited number of movable parts.

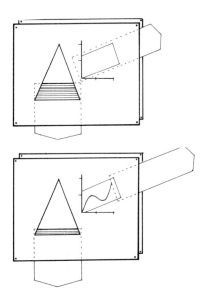

With superimposition or inlay, animated captions can be incorporated into special effects sequences. A moving graph, for example, could be superimposed over a scientific experiment in

The sequence involved in three-tone
animation.

order to provide students with an easily perceived measure of the 'background' events. High quality *three-tone* animation can be achieved in monochrome by superimposing over an even grey background, one white on black animated caption, simultaneously routing a second white on black animated caption through inlay, and changing the white information to black. Combinations of this sort permit sequences of some sophistication, but also require very accurate construction (allowing perfect registration between elements which are actually contained on different captions), and a great deal of studio rehearsal time to enable cameramen, caption operators, and the verbal commentator to co-ordinate their efforts. As caption sequences become more complicated, the temptation to substitute animated film (if that is available) increases.

Film animation   This technique encompasses such a wide range of skills and theory that any brief account can do no more than outline basic production methods and suggest some of the uses to which they can be put.

The illusion of movement can be achieved with cine film because of a characteristic associated with human visual perception, usually termed 'the persistence of vision'. The retina of the eye will retain or 'store' the image of an object for a fraction of a second after the object has disappeared. Film capitalizes on this physiological function by rapidly replacing the picture projected from one frame by a series of slightly different pictures on successive frames. If frame changes are made in a shorter period than that of the persistence of vision, a viewer experiences the impression of movement.

Film images of actual events are recorded continuously by a camera at a rate of 24 frames per second (if it uses 35 mm or 16 mm stock). After laboratory processing one has to examine a series of frames to understand how slight variations between pictures accumulate to produce the appearance of movement when they are projected. Film animation uses exactly the same principles but, because there are no 'real' events to record, filming cannot be continuous. Each successive frame must still carry a slightly different image, but this can only be achieved by changing or moving pictorial elements in periods *between the exposure of individual frames*.

This basic principle can be applied in a number of ways, though the degree of complexity possible in the finished result will be limited by the sophistication of the film equipment used.

For simple animation effects the only essential piece of equipment which needs to be purchased is a film camera with a single-frame exposure control. While cameras for live action cinematography are motorized to transport film stock continuously past the lens at an even rate, those for animation sequences must be capable of presenting a single frame of film stock for exposure in exactly the same registration as previous and subsequent frames. In other words, there can be none of the small variations in the framing of a scene which sometimes appears in actuality filming and is known as 'picture bounce'. Other ancillary equipment can be built by the small ETV organization.

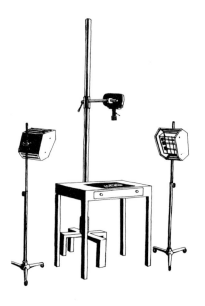

A basic animation unit might comprise a single frame camera, a stand to hold the camera rigid in different vertical positions, a table on which to position artwork beneath the camera, and lights which will provide illumination of the same intensity over hours or even days of filming.

With such a unit, the designer can produce animated sequences using the following techniques:
(a) With cut-out drawings or abstract shapes which are moved about on a background to a pre-arranged plan. The cut-outs are filmed frame by frame, in a series of slightly different positions; smooth movement, for example of dots (maybe symbolizing atomic molecules) travelling across the screen, can be achieved if the space between each position filmed is kept small and even. The designed speed of animation can be effected by allocating to any particular

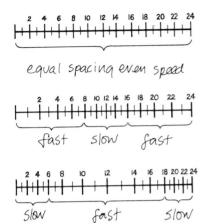

equal spacing even speed

fast   slow   fast

slow   fast   slow

movement a larger or smaller number of single-frame exposures. A movement allocated twenty-four frames will occupy one second of screen time when projected, one of 144 frames will take six seconds – and so on. When the frame allocation has been made, the designer must describe an exact position for the cut-out on each frame. Artwork for cut-outs must be prepared on thin card as paper will curl under the heat of the lights. It may even be preferable to use magnetized cut-outs (see page 52) on a tin-plate background, since the pieces under the camera are then unlikely to be accidently knocked out of position or blown away by a strong breeze!

(b) By drawing under the camera, so that the animation is actually built up by sketching in successive parts of the diagram during filming. A faint guide (invisible to the camera) can be traced out of the completed picture, and a frame allocation made according to the speed with which each new line of the diagram is to appear. The designer will then draw in the first frame of the diagram and film it, draw the second frame and film, and so on.

(c) By using flat models. These resemble studio-animations (see page 53) though they will have been carefully calibrated to unmask the pictorial elements according to a frame-by-frame plan. This technique is very useful when precise but repeatable movements, such as the mechanical movements in part of an engine, are required. To avoid the windows and masks of the model appearing on film, special high contrast black-and-white stock is used, which does not register the grey scale variations produced by shadows.

When a designer is required to produce more complex sequences (e.g. animation with a moving or changing background, or with different elements animating in different ways at the same time) he will prepare his artwork on a series of cels. These are clear acetate sheets which are usually punched with three holes down one edge to fit over registration peg bars on the animation table. With this positioning device layers of cels can be built up to compose one or more moving shapes, and these can be positioned exactly in relation to a changing background. The basic animation equipment has to be supplemented with mechanical features to accommodate cel artwork.

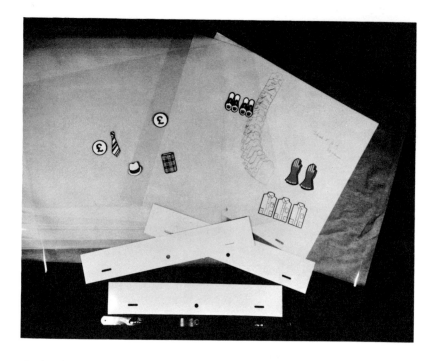

Layers of cels (clear acetate sheets) can be accurately registered by means of the three holes fitting over peg-bars; part of the technique to achieve moving shapes.

A simple cel animation unit will comprise a single frame camera, a camera stand, an animation table which will move beneath the camera as a single whole or permit differential movement between cel layers, and a glass platen to compress the cels flat when filming. Polarized lighting will also be needed to avoid reflections from the cels.

The designer will now be able to use a wide range of animation techniques, the principal constraining factors being the quantities of time and effort that can be afforded in the planning and production of cel artwork. In fact even handling the cels requires great care since they mark easily, and gloves must therefore be worn.

The outline of any shape drawn on cel has to be filled in by painting *on the back of the cel* to ensure that an even flat colour is presented to the camera. Movements have to be planned so that the same number of cels are in use throughout a sequence. Otherwise variation in the density of the filmed image can be expected.

Cel animation utilizes the same fundamental principles as the cut-out technique described above, but now, instead of moving a single cardboard shape between exposures, a completely different cel is positioned for each new film frame. Successive cels contain representations of the same shape in slightly different positions. Because the transparent cels are kept in registration by peg bars, opaque backgrounds can be changed without affecting foreground shapes and their movement. Artwork is drawn on different cel layers because some pictorial elements will probably need to remain motionless during some part of the sequence. In this case, their cel will be left unchanged during exposures until they are to resume their animation.

A simple but effective technique for making lines 'grow' without elaborate artwork preparation is the *scrape back* method. A *complete* diagram is painted on a cel, positioned on the animation table and filmed with the film stock travelling *in reverse*. Selected parts of the artwork are successively scraped off the cel between exposures, so that when the processed film is projected forwards, the artwork will appear to develop to its fullest extent.

When an organization is involved in the production of long and complicated animation sequences, it will probably consider purchasing additional equipment which can economize on an operator's time by making certain effects easier to film.

A complex animation unit would, for example, include a specialist animation camera, incorporating facilities for fade-ins, fade-outs, and dissolves; it would also have precision mechanisms for reverse and forward filming, and automatic lens focus. The camera will be mounted on a carriage which rides up and down on calibrated vertical columns which are secured to the rear of the animation table. This movable carriage enables the camera to be conveniently placed

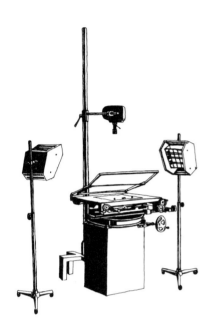

A simple cel animation unit – single frame camera, camera stand, animation table, glass platen, and polarized lighting.

A more complex animation unit,
including a specialist camera mounted
on a carriage.

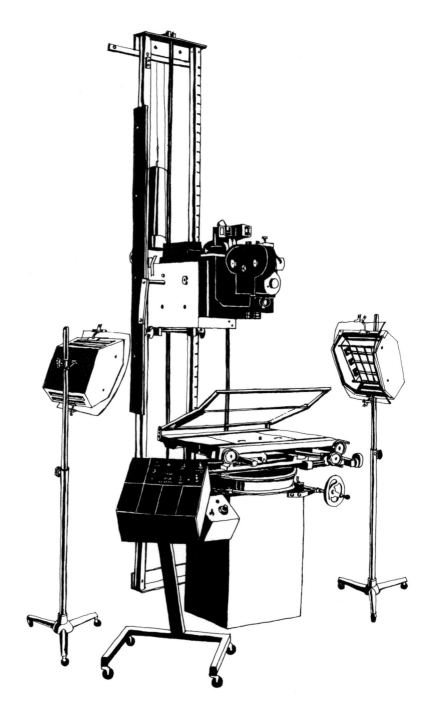

at any distance from the artwork that the required film area positions may demand, and permits precision 'zooms' to be achieved. A control panel may be added to the animation unit to regulate both the exposure of the film and its movement through the camera. Operating time is thus reduced because the camera is partly controlled through motorized switches.

This sophisticated equipment will enable the designer to produce sequences in which different pictorial elements fade in or out, and where dissolves between scenes can be made. Precise registration mechanisms in the camera will also permit successful filming of *multi-exposure* sequences. Here some elements of a diagram are exposed on the film's first passage through the camera; then the film is wound back (with the shutter closed) to the original start position so that additional elements can be exposed during a second passage. Artwork for the second exposure is often prepared white on a black background; the white areas will be visible on the final picture, while the black sections protect the previous scene from obliteration.

Full animation of cartoon figures requires this type of expensive equipment. Fortunately, perhaps, this simulated 'naturalism' is rarely needed in ETV. Other and cheaper methods like the ones mentioned above can give results which are as good or even better for educational purposes.

The production process Because mistakes often prove very expensive in film animation, designers must take every reasonable precaution to ensure that directors and teachers have approved their plans before filming begins. After initial consultations about the function and treatment of a sequence, the designer will prepare a detailed storyboard which embodies sketches outlining the proposed movements and changes of scene, together with any related commentary, music, or sound effects. When this has been approved, the designer can use it as a basic plan and reference for the information of all those involved in compiling the animation.

If the animation is to be matched precisely with a sound-track, the latter will be pre-recorded onto magnetic film sound stock. Using film-editing equipment, the designer can then compute at exactly which picture frames certain movements must begin and end if they are to synchronize with specific events on the sound-track. This information is incorporated into a timing sheet (dope sheet), along with accurate frame-by-frame instructions regarding all other timings, e.g. the movement of the artwork on the animation table, camera moves in relation to the artwork, dissolves, multiple exposures, etc.

Meanwhile artwork for each sequence will have been in preparation,

Timing sheet (dope sheet).

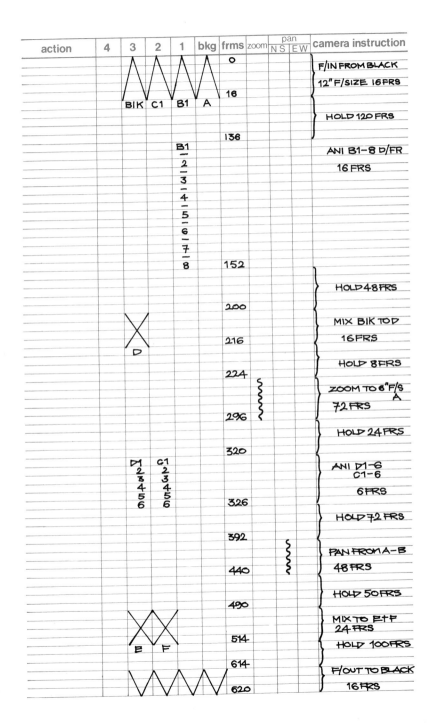

action	4	3	2	1	bkg	frms	zoom	pan N S E W	camera instruction
						0			F/IN FROM BLACK
									12" F/SIZE 16 FRS
		BIK	C1	B1	A	16			
									HOLD 120 FRS
						136			
				B1					ANI B1-8 D/FR
				2					16 FRS
				3					
				4					
				5					
				6					
				7					
				8		152			
									HOLD 48 FRS
						200			
			D			216			MIX BIK TO D 16 FRS
									HOLD 8 FRS
						224			ZOOM TO 6"F/S A 72 FRS
						296			HOLD 24 FRS
						320			
		D1	C1						ANI D1-6 C1-6
		2	2						6 FRS
		3	3						
		4	4						
		5	5						
		6	6			326			
									HOLD 72 FRS
						392			
									PAN FROM A-B 48 FRS
						440			
									HOLD 50 FRS
						490			
		E	F						MIX TO E+F 24 FRS
						514			HOLD 100 FRS
						614			F/OUT TO BLACK
						620			16 FRS

and when completed should be checked by the director. Any minor errors must be discovered at this stage if costly re-shoots are to be avoided.

The sequence is then filmed on the animation unit and if there is time beforehand a short test film will be exposed and processed before the main shooting begins, in order to check the pace of movements.

It is always advisable to schedule time for re-shooting, since quite apart from human fallibility, faults in the camera, film stock, or processing may otherwise extend production time beyond the deadline.

Chapter 5 **Design in action**

Although organizations which make television programmes can be very large indeed (the BBC, for example, had 25,000 employees in 1972), programme-makers themselves tend to work in small groups. The size, composition and life-span of a *production team* will depend on the type of programming involved and the degree of specialization available within the parent institution. In its most developed form, an ETV team will probably include a producer, a director, an academic scriptwriter/presenter, a set designer, a film crew, a film editor, a producer's assistant and secretaries, as well as a graphic designer. Obviously many other specialized and important skills are involved in producing a programme (e.g. TV cameraman, videotape engineers), but where this effort is relatively short-term and not directly concerned with the planning of programme *content*, the expertise is not regarded as part of the production team itself.

A production team is headed by a producer (or sometimes a director) who is responsible for planning and budgeting a programme – or, more often, a series of programmes – and also for co-ordinating the work of other team members. As the person held responsible for achieving a high-quality result within a specified budget, the producer has the right to make final decisions in all areas which affect the team – though an experienced producer will seek the advice and opinion of his co-workers who are more skilled than he in their individual specialisms. In ETV, a producer is likely to be either a qualified teacher who has acquired knowledge of television production, or an expert in production who confines himself to the presentation and interpretation of material which has been originated by an academic adviser (who is sometimes also the presenter). But as the importance of ETV is increasingly recognized, a new breed of producer is beginning to appear who combines both sets of qualifications.

The production team is headed by the producer who is responsible for planning and budgeting a programme.

Whether or not the presenter is also the originator of a programme's educational content, his is a crucial role, for the effectiveness of every other programme component – including graphic materials – is to some extent dependent on his ability to maintain student attention and comprehension. If he is merely a performer giving expression to ideas generated by someone else, the designer is unlikely to have much contact with him, except during studio rehearsals, when last-minute modifications may be required to graphics which the presenter personally manipulates. But when the presenter is also the author of his script, it is advisable for a designer to establish a rapport with him and his ideas, so that three-way meetings (i.e. between presenter, producer, and designer) can be conducted efficiently and without the sort of time-wasting misunderstandings that arise so easily when decisions result from group discussion.

A set designer will work closely with the producer and director from

the original conception stages of programme preparation. His
primary task is to provide an environment in which the disparate
elements of content can be brought together in a convenient,
coherent and attractive manner. Since the graphic designer's work
will be displayed somewhere in the area for which the set designer is
responsible – whether it be a specially designed display board
located within the set itself, or an ordinary caption stand placed in
an unused corner – discussion between the two is essential
throughout the planning stages. This co-operation will extend
beyond the strictly practical concern of how best to present graphics
for the TV cameras to the more difficult problems of establishing a
consistent style between their creations.

ETV programmes often need to record people, events or things
which cannot be brought into an electronic TV studio. In this case,
film shot on location is inserted (via a telecine machine – see page
37) into TV studio sequences during video-tape recording. A film

cameraman, a sound recordist, and their assistants may therefore work closely with a director for several days, shooting material for inclusion in a programme which will probably not be finally assembled for several-weeks. If a film crew is to record exactly what the director wants, it is essential that its members be regarded as creative partners who must fully understand the educational objectives of their work before they begin shooting. The same is true of the film editor, who assembles the filmed 'rushes' sent back from location into coherent sequences ready for telecine projection. The editor also works closely under the director's supervision, but will also have contact with the designers, particularly when he is cutting filmed photographs and diagrams or animated film.

A producer's assistant – or sometimes a secretary – will attend to the paperwork connected with the producer's co-ordinating activities. Designers are well advised to maintain clear channels of communication with producers' assistants and thus to ensure, for

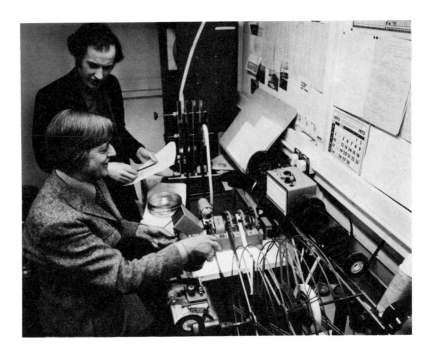

example, that design decisions which have been agreed with
producers and directors are confirmed in typewritten memoranda.

The combined efforts of a production team require co-ordination so
that one member's finished work is ready whenever it is needed by
a colleague. Like other team members, a designer must observe
several pre-studio deadlines for completion of various types of work.
Moreover, since most organizations believe that a graphic
designer's efforts are most economically deployed if he works
concurrently on several programmes, and since this implies working
towards a series of different deadlines, it is imperative that
administrative procedures are developed that will keep the designer
constantly alert to which assignment must be completed by which
dates.

The first stage in the production process for a designer will be his
initial meetings with the producer or director, at which different

possibilities for presenting information will be discussed, along with some of the practical constraints (budget, available skills, viewing conditions, etc). As with other design fields, the more explicit the original brief, the more likely is it that the designer can produce work which fulfils its function. Needless to say, it is partly the designer's responsibility to ensure that he has received an adequate account of what is expected of him; and only then can he confidently begin his own creative contribution. On page 72 there is a checklist of the minimum amount of information that a designer should secure before starting work on a programme.

If film animation is to be used, planning and production (see page 61), must begin immediately after first consultations with the director. The film editor will supply dates by which he must receive the processed film. If it is going to be necessary for him to make cuts in the film animation or its soundtrack, his deadline will be earlier; if the animation is to be shot to an exact length (which has probably been computed from a pre-recorded soundtrack – see page 62), the editor has only to join it into the other film sequences he has been cutting – in which case he will not expect to receive it quite so quickly. It is usually worthwhile for designers to discuss animations with the film editor before compiling the final dope sheet since his advice may assist them in planning for smooth cutting or in avoiding possible snags.

Occasionally graphic materials will be required early on in a production schedule, for example during location filming. Scientific apparatus may need labelling or an expert contributor who cannot make the journey to the television studio may need diagrams for the pre-recorded explanation he makes on film. Although the designer will probably not attend filming sessions, he must have the appropriate graphics ready for the director to take with him on location.

Where large-scale studio graphics are to be used in a programme, conversations with the set designer will aim at integration of design styles but will also cover such pragmatic topics as the sizes of graphic elements in relation to their position within the set and the camera angles planned by the director, the tonal treatment to be given to different areas of display, and the types of surfaces to be used. Some large graphic displays require a lengthy preparation time – maybe outside contractors will be involved – and then planning and execution must begin early. The same is true of other graphic elements which, like studio animations, cannot be changed easily and utilize outside skills. If possible, design schedules should be compiled to allow time for reconstructing graphics where errors have been made.

Artwork that has to go through a photographic process (e.g. slides and photo-captions) is the next priority in most schedules, and where video-animations are to be made before the main studio day, artwork is likely to be required about the same time. Meanwhile typographic printed material should be under way, and the spelling and layout checked on completion.

Finally the small-scale studio graphics (e.g. studio captions, labels, etc.) will be started. These are comparatively easy to modify, and changes do not involve the labour of anyone but the designer. All studio graphics should be ready at least twenty-four hours before the start of the recording day. As practising designers know, there are likely to be enough changes required during the studio rehearsals to keep them busy without the necessity of rectifying errors which have slipped through unchecked.

During the studio day the designer should be on hand to effect any last minute changes, and even to produce new graphics should the need arise. It should be said, however, that if too many additions or adjustments are necessary at this stage, either the director or the designer has erred in failing to anticipate the proper requirements. Rehearsal time spent in making elaborate alterations to studio graphics is the waste of a costly facility (the studio).

Checklist of information

After reading the programme script or outline, the designer should ascertain the main educational objectives of the programme and the crucial teaching points and then proceed to gather the sort of information listed below.

1 The nature of the audience
  (a) age group
  (b) educational level
2 The nature of the audience's viewing conditions
  (a) domestic viewing
  (b) classroom viewing
  (c) large lecture theatre viewing
  (d) the television receiver sizes for (a) (b) (c)
3 The graphics budget for the programme
  (a) resources/money
  (b) time
4 The production treatment
  (a) graphics with the presenter
  (b) graphics without presenter
5 Graphics with the presenter
  (a) display boards – size/camera field areas
  (b) back projection – size of screen/camera field areas
  (c) colour separation overlay – related colour of background and foreground/field areas of foreground, background
  (d) graphics or models – camera angles, field areas
6 Graphics without the presenter
  (a) captions – field areas
  (b) studio animations – sequential order and type of movements
  (c) telejector slides – field areas/if superimposed, nature of background
  (d) film/video-tape animation – availability of sound track recording/pre-studio deadlines

A design schedule.

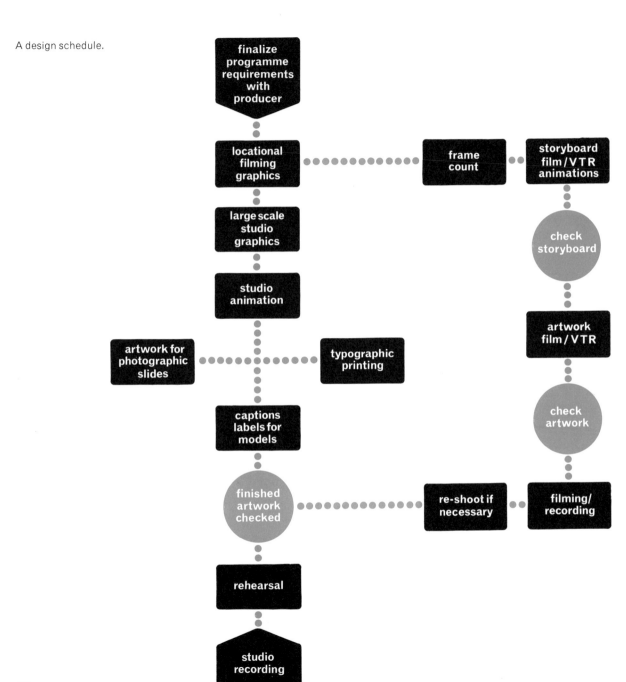

finalize programme requirements with producer

locational filming graphics

frame count

storyboard film / VTR animations

large scale studio graphics

check storyboard

studio animation

artwork film / VTR

artwork for photographic slides

typographic printing

check artwork

captions labels for models

finished artwork checked

re-shoot if necessary

filming/ recording

rehearsal

studio recording

When the recording has been made (if not before) certain administrative details must be transacted by the designer, probably in conjunction with the producer's assistant. Storage of graphic materials must be arranged, if it is the organization's policy to do so; final costings of the graphics must be provided, so that production staff can complete the programme account and if any graphics have incorporated copyright material, details must be furnished so that the necessary clearance can be obtained.

Obviously this pattern of working is not invariable. Sometimes, for example, the artwork for small captions may be more complicated and time-consuming than that for studio animations or even film animations. The description given above represents only a kind of norm; in practice many programme schedules will differ to a larger of lesser extent. Part of an ETV designer's expertise lies in the initial assessment of the time which needs to be allocated to the preparation of each item required, and in the scheduling of effort in such a way that each component is ready when needed.

Research and ETV   Surprisingly, there is little academic research currently available which will help the ETV designer improve his effectiveness. Media research has characteristically concentrated on the effects of individual programmes, series, or overall features such as violence and sex on TV. Research into educational television has been conducted within this general perspective – and at the moment is facing something of a crisis, being unable to provide *generalized* statements which will guide programme-makers. The main problem lies in the difficulty of experimentally controlling the large number of variables present in any teaching situation. In other words, although we can discover with reasonable accuracy whether an individual programme has achieved its educational objectives, we can never be certain *why* it has succeeded or failed: any number of possible reasons can be envisaged (teachers' performance, graphic style, subject matter, pace, student capacity, and so on), but until we can methodologically isolate the influential factors, it is impossible to be sure which elements of our programming are having what effects.

It now seems clear, therefore, that if research is to benefit the ETV designer – or any other member of the production team – it must set itself three principal tasks:
to evolve a terminology which will reliably describe the audio and visual signs and symbols through which programme content is expressed;
to determine which sorts of signs and symbols television can convey;
to explore how certain concepts can be embodied in these signs.

Like every other working professional, the designer would like to have available general rules which could tell him how to proceed in any assignment. These do not exist, except in the case of legibility studies mentioned below. With regard to general graphic styles our intuition must play a larger part than may be desirable, for although it is possible accurately to investigate the effectiveness of a single programme in communicating a given number of ideas, we cannot be sure why this is so.

Such research as has been published seems inconclusive when dealing with graphic materials. This may not be surprising when we consider that no adequate language has been developed to classify graphical items. If we cannot describe programme elements in a consistent manner, it is no wonder that we are unable to assess their effect on student audiences.

Even though many of a designer's stylistic choices continue to be matters for personal judgement, legibility studies have provided clear evidence about the minimum sizes of pictorial components on the TV screen in relation to maximum viewing distances. The size of screen used to display television pictures largely determines viewing distances; these have been established as twelve times the width of the particular screen, so that, for example, the maximum

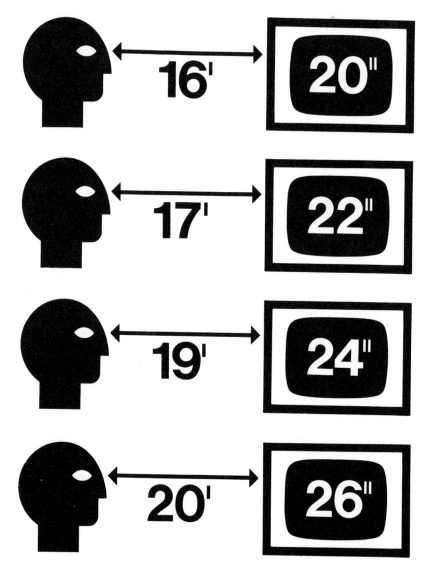

distance possible for pictures viewed on a 24 in. receiver (for which the dimensions are 19·2 × 14·4 in. as shown in the diagram on page 29) will be 19 feet. This calculation assumes that the minimum height of individual pictorial elements will be no less than 1/25 of the total picture height of the television screen.

In the pages which follow, examples are given of some of the work being undertaken by designers in various television stations in the UK, USA, and Japan. They give only a brief glimpse of the kind of programmes being transmitted by the stations and networks concerned, but when the stills are read in conjunction with the accompanying notes, some idea can be gained of the variety of approach and of the design possibilities being exploited to meet differing educational requirements.

Among the examples shown, the University of Leeds caters on a closed-circuit basis for an essentially academic audience; the BBC Open University programmes are aimed at a similar type of audience, but on a broadcast basis. The ILEA programme is an example of a closed-circuit production aimed at junior children; the examples from Maryland and NHK Japan are also devised for younger children, but are broadcast programmes. The item from ATV demonstrates the instructional possibilities for technical broadcasts, and, finally, the contribution from the Children's Television Workshop in New York provides an example from a programme in a series designed to reach out to the educationally under-privileged.

University of Leeds, England

The University of Leeds Television service was set up, staffed, and equipped to provide a whole range of closed-circuit and audio-visual techniques. From its early days in 1964 when it operated out of a converted warehouse, the service has now expanded to a staff of twenty-seven accommodated in a purpose-designed building.

A graphics unit, which was created to service the television studios and film unit, now engages four designers in a great variety of work from television to complicated animated film sequences.

These stills are from a series of three paleographic films produced by the Television Service and the School of History. The films were made for postgraduates, librarians, and archivists studying mediaeval records to aid understanding and hasten the process of textual transcription of documents.

The problems for the designer were twofold. First, to reproduce the documents photographically, either from the originals or facsimiles in a suitable form for 16 mm black-and-white animated film; and second, in view of the highly detailed nature of the scripts, to preserve the complex letter shapes and fine hair lines produced by the quill on parchment.

To reproduce the documents effectively it was necessary to achieve a high standard of photography in order to restore the contrast between script and parchment without losing any of the original detail.

One of the film's main aims was to decipher recognizable letter shapes. It was important to study the letter forms and set down a logical evolutionary pattern, which was demonstrated by writing, with a quill, the alphabets of three types of scripts (rustic, uncial, carolingian) showing the angle of the quill while writing and its curving characteristics resulting in the individual letter shapes.

In each film a section was devoted to text transcription. Lines of text were lifted from the manuscript, then filmed with the transcription (in a roman type face), animating word by word underneath each line.

1

4

7

2

	A B C D E F G
rustic	A B C D F̄ G
uncial	λ B C d e F G
caroline	a b c d e ſ з
	α

5

simulobdormier̄necresurgent :
simul obdor mierunt necre surgent
contritisuntquasilinum et
contritisunt quasilin um
extinceisunt; nomemineritis

priorum etantiquaneintueamini ;

8

3

6

simulobdormier̄necresurgent :
simul obdor mierunt necre surgent
contritisuntquasilinum et
contritisunt qua silinum et
extinceisunt; nomemineritis
extinctisunt ne meminer itis
priorum etantiquaneintueamini ;
prior um etantiqua neintueamini

9

ILEA, London, England

The ILEA Television Service was set up by the Inner London Education Authority in 1966 to serve the schools, colleges, and other establishments maintained or aided by the Authority. These number nearly 1,400, and television programmes covering virtually every subject at varying levels are distributed via a closed circuit coaxial cable rented from the Post Office. Production is in the hands of London teachers, who receive in-service training as researchers, script-writers, presenters, and studio directors. They are supported by the normal range of television facilities and resources, including a design department which is divided into two groups, graphics and sets. On the graphics side, the work is shared by three designers, with two assistants, responsible for programme illustration through captions, diagrams, animations, etc., and also for illustrative ancillary literature. Set designs are the responsibility of a further three designers, with one assistant, supported by construction, props and painting sections. In all, sixteen design staff service the needs of two studios, producing around 300 programmes in a year, and a smaller training studio.

The stills opposite are from a mathematics series which was primarily intended for final year junior pupils aged around 10. Dealing with the history of number, it was scripted with the dual aim of imparting a certain amount of information about number devices and patterns and presenting this information in a stimulating and enjoyable way. The presenter or television teacher is therefore embodied as a young 'professor' who conducts his researches into the historical development of number by materializing at various points in time with the aid of an idiosyncratic, but omniscient, time machine called Roton. Each historical expedition is presented in dramatic form, and the settings are designed to supply not simply a period backcloth but a vehicle for the natural and convincing presentation of mathematical teaching points. For example, the tiled walls of a Babylon set embody patterns demonstrating the properties of square numbers, and the Centurion stationed on Hadrian's Wall (see illustration opposite) explains Roman numbers by scratching them – *graffiti* style – on the wall of the fort.

1

2

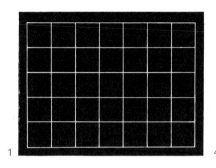

3

4

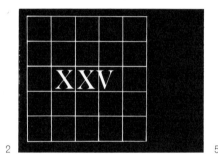

5

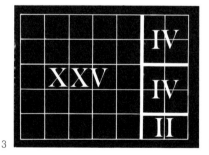

6

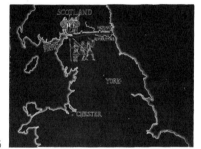

7

8

9

Maryland Public Television, USA

The Maryland Center for Public Broadcasting is the headquarters and production facility for the Maryland Public Television network.

The production facilities include three colour television studios, colour telecine machines, an outside broadcast unit, and a film unit.

The graphics staff totals twenty-seven, producing titles, promotion, and content support; printed graphics – promotion and programme support material; animation; still photography; and scenic design. Though they operate in specialized areas, the philosophy is to operate as a team so that all elements of a production – scenic, animation, graphics, photography, and print develop coherently, thus increasing the overall effectiveness and impact of a programme or series.

In the 'numbers game' illustrated opposite there were six concepts in mathematics for students aged 7–10. In order to attract and hold the attention of the students, the producer and designers developed the concept of a game. The game included a game board and pieces similar to chess men. The design of each chess piece was based on a geometric shape and represented a particular concept. It was through this device that the television teacher introduced each learning segment. This game was used in conjunction with a mini-set which

solved other problems, both graphically and scenically.

The graphics for each programme were quite varied. Styrofoam geometric shapes were used to show division, i.e., a circle would be pre-cut into quarters and then reglued with rubber cement. This would be 'broken' by the teacher to show the division. At other times a similar device would be used but would be attached to a magnet board. By using tape editing, the pieces could be moved on the magnet board to achieve an animated effect.

In order to solve other graphic problems, it was decided to animate certain geometric shapes and numerals but in a very simple manner. This was accomplished by attaching brightly coloured sticks to black rods. These were manipulated against a black background to form a geometric form. Foam rubber numerals were animated by the insertion of black rods with the numeral activated against black. All graphics were kept as bold and clean as possible.

1

4

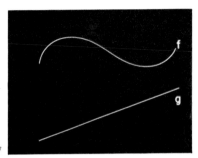

7

2

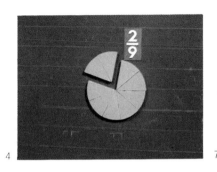

5

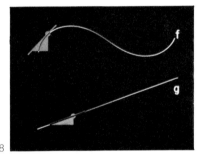

8

3

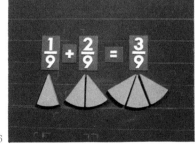

6

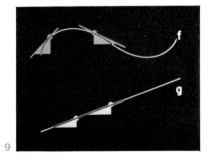

9

NHK Tokyo, Japan

1

2

3

4

5

6

NHK operates seventy-one broadcasting stations throughout Japan and provides services with two networks both in radio and television. Similar in character to most European broadcasting organizations, it is a public service broadcaster and operates on a nationwide scale.

NHK attaches the greatest importance to educational programmes; its second radio and television networks are devoted exclusively to education, transmitting from 6 am till midnight.

Graphic design normally required for NHK's television programmes, i.e., programme titles and maps and charts, illustrations, etc., are supplied almost exclusively by the NHK Art Center, which employs more than a dozen full-time graphic designers.

In addition to these designers freelances are also commissioned to execute designs from time to time as need arises. Leading designers in Japan are frequently seen taking part in the planning and design of programme titles, illustrations, and animations.

This storyboard is part of an animated film *A Mini Symphony,* one of several films in a series called *Animated Fantasies* for small children. The series is intended for repeated use over television to help children cultivate a sense of beauty, and develop their creative powers by arousing their interest in all sorts of forms and colours.

7

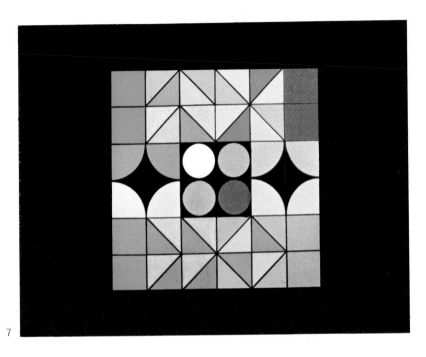

85

ATV, England

1

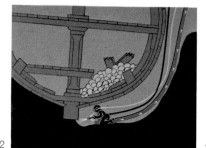

2

3

ATV's studios are in two locations, the main transmission complex at ATV Centre, Birmingham, and a large production unit at Borehamwood, Herts, combining all the facilities necessary for a colour television service at broadcast level.

This storyboard is part of a film animation sequence in a programme from a series called *Secrets of the Deep*. It illustrates the techniques of a salvage operation in 1960, to raise a Swedish flagship, the *Vasa*, which capsized on her maiden voyage in 1628.

(1) Divers tunnelled under the delicate hull of the old ship.
(2) Used a high pressure hose to free the mud which was sucked up into a return pipe.
(3) Cables were put around the hull and made fast to ...
(4) Partially flooded pontoon tanks at the surface.
(5) Air was pumped into the tanks ... and down below ...
(6) The hulk of the *Vasa* lifted clear of the mud.

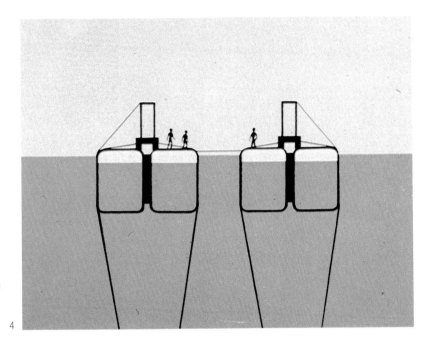

4

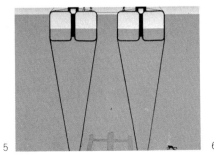

5

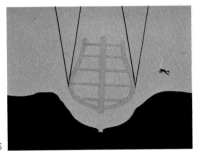

6

BBC Open University, England

Open University television programmes were first transmitted on BBC2 Network in January 1971, the start of an educational partnership between the BBC and the Open University which has since been labelled 'the most advanced and sophisticated multi-media instructional system to teach large numbers of students at a distance'.

The graphic design unit grew out of the parent department of BBC Television. Obviously, new attitudes and skills were needed to cater for the shift in emphasis from presentational to informational graphics. The need for a more integrated design approach than was previously possible in educational programmes was apparent.

Current programme-making for the Open University consists of six main faculty areas: arts, social science, education, science, technology, and mathematics. Designers and their assistants are assigned to courses or series of programmes within each faculty, working with production directors and academics who are specialists in their own subject areas.

Now, after more than two years of concentrated effort, there appears to be emerging an awareness, particularly in the academic areas, that the visual is not merely an aid to verbal instruction, it is a language in its own right, and fluency in that language must be a prerequisite to effective communication.

1

Here is an example of one of several multiplane film animation sequences in a programme for the arts faculty dealing with the iconography of song and dance in the seventeenth century. The lavish theatrical spectacles staged in Florence during that period were uniquely recreated by using original stage and costume designs, and accounts by contemporary authors.

Children's Television Workshop, USA

The Children's Television Workshop is a non-profit organization, supported with funds from United States government agencies and private foundations. Its board of advisers is composed of some of the most outstanding authorities on education, psychology, and other disciplines associated with learning in the United States. The workshop currently produces two programmes *Sesame Street*, and a new production called *The Electric Company*.

Sesame Street, originally designed to reach poor children in both urban and rural areas of the United States – many of whom are members of racial and ethnic minorities – was most thoroughly researched before it went on the air. And because it is considered an experiment, research continued throughout the first season of broadcasts.

Sesame Street incorporates entertaining variety and fast pace – which Workshop researchers have found appealing to young children – with an educational curriculum to provide several basic skills for youngsters before they enter school. These skills include learning the English alphabet and other preparation for reading, the numbers from one to ten and geometric shapes, exercises in reasoning and problem-solving and developing an awareness of self and the world around us.

This example is a part of a film animation teaching children the number three.

1

2

3

4

5

6

7

8

10

9

11

12

Glossary

animated captions
Constructed from layers of card, operated manually, they involve the uncovering or concealment of information at different points in time.

animation
Techniques of animation are employed in ETV when it is necessary to demonstrate processes by showing pictorial elements in dynamic inter-relation. Diagrammatic movement can be used to highlight changes in size, speed, or density; to illustrate direction or flow patterns; to delineate sequential patterns of a spatial or temporal kind; or merely to direct the viewer's attention to different parts of a diagram. The three types of animation in most common use by ETV producers are magnetic diagrams, animated captions, and film animation.

aspect ratio
Refers to the horizontal and vertical dimensions of the television screen. The width to height ratio is always 4 : 3 on modern receivers. Screen sizes are measured along a diagonal line from corner to corner.

back projection (BP)
A high-powered projector throws an image of a photographic transparency onto a translucent screen. The television camera views the screen from the other side, and a presenter is able to appear in the same area.

broadcast TV
Programmes broadcast via transmitters, so that any TV set with suitable aerial and within range can obtain sound and picture images.

cel animation unit
Suitable for complex animation sequences, involving a moving or changing background, or different elements animating in different ways at the same time. Clear acetate sheets, with three holes down one side for accurate positioning over *registration peg bars* (see below), are used to build up moving shapes.

closed-circuit TV
Programmes carried to a limited number of TV receivers by way of coaxial cables or a microwave link.

colour separation overlay (CSO)
A foreground subject is placed in front of a highly saturated blue background; this colour activates a keying switch in the *overlay* device (see below) which results in the blue area being replaced by pictures from a second source. Produces higher quality composite scenes than *back projection* (see above), and is useful in small studios where a presenter in a single location can be shown in association with an unlimited series of high-definition pictures, still and animated. Also known as chroma-key.

cut-aways
Visuals shown without a presenter.

cut-off
The capacity of TV receivers to display the total picture transmitted varies considerably. Producers and designers must allow for a 'cut-off' area to safeguard against picture loss in transmission. This may be as much as 20% of the picture area and a safety margin must be left around artwork and photographs. Vital information must not appear in the cut-off area.

cut outs
Animated sequences can be produced by moving cut-out drawings or shapes on a background to a pre-arranged plan. The cut-outs are filmed frame by frame in slightly different positions. Sometimes the cut-outs are magnetized so that they adhere to tinplate display boards.

dope sheet
A timing sheet on which the designer incorporates accurate frame-by-frame instructions concerning movements of artwork on the animation table, camera moves, dissolves, etc., so that they synchronize with the soundtrack.

film-editing equipment
The designer can use this to compute at exactly which picture frames certain movements must begin and end to synchronize with the soundtrack.

flat models
Carefully calibrated models which enable the pictorial elements to be unmasked according to a frame-by-frame plan. A useful animation technique to demonstrate precise but repeatable movements, e.g. parts of an engine.

glass platen
For compressing cels flat when filming.

grey scale
The range of tones between black and white which TV cameras and receivers are capable of reproducing. A high-quality system can distinguish as many as twenty shades, but the range will be reduced in transmission, and artwork for educational TV often makes use of as few as five tones, to ensure clarity of communication. Artwork for colour programmes must still relate to distinguishable tones on the grey scale to allow for monochrome receivers.

hot-press machine
Equipment for producing captions. Heated lead type is pressed onto thin plastic foil which covers the caption card. Under pressure, ink is released from the foil and leaves an impression of the type.

inlay
A method by which part of one television picture is electronically cut out and replaced by visual material from another source. Useful in educational TV for achieving 'split-screen' effects (see below) so that information from different sources can be compared.

letterpress transfer
Individual letters on thin plastic sheets which can be transferred to artwork by rubbing. Time-consuming, and only suitable for small quantities of lettering.

line length
In determining maximum length of a line of type which can appear on the screen, designers must take account of factors such as legibility and cut off. On 12×9 in. artwork the maximum length of line is 8½ in.

line structure
A television picture is made up of fine, horizontal lines. According to the system in use, which varies from country to country, there may be 405, 525, 625, or 819 lines per picture. Other things being equal, the more lines, the better the picture resolution (i.e. ability to reproduce detail).

magnetic tape
Used for recording sound and picture signals on a video-tape recorder (see below). Varies in width between ¼ in.–2 in. according to picture quality required. Can be erased and re-used.

multi-exposure sequence
Precise registration mechanisms in a camera permit such sequences. Elements of a diagram are exposed on the film's first passage through the camera; then the film is wound back (with shutter closed) to the start position so that additional elements can be exposed.

overlay
Works on a similar electronic principle to inlay (see above). A foreground subject from one camera can be inserted into a background picture from a second source (camera, slide, or telecine). The special version of this for programmes in colour is termed colour separation overlay (see above).

panning
A continuous horizontal movement either of a camera across artwork or artwork in front of a stationary camera.

photo blow-up
With monochrome systems, a photographic enlargement can be prepared for studio display. The cost and quality of colour blow-ups prohibits their use. With colour systems, projected photographic images are used instead.

photo-typesetting
Where special typographic effects are needed (such as the distortion of lettering), characters stored on film and printed photographically can be used to produce the required forms. Usually considered too expensive for more than occasional use in educational TV.

picture bounce
Small variations in the framing of a scene in actuality filming.

picture resolution
The ability to reproduce detail, largely determined by line structure (see above).

playback
The reproduction of recorded material (vision and sound), through receivers linked to a video-tape recorder (see below).

polarized lighting
Used with a cel animation unit to avoid reflection from the cels.

presenter
The person, often a teacher, who either appears with visuals to point to and manipulate diagrams, or describes them off-screen, to avoid student distraction.

projected display
Artwork for projected display will need a high-quality finish, because blemishes will be magnified, and most studio projection devices involve picture quality loss, which needs to be compensated for.

registration peg bars
To secure perfect alignment of cel acetate layers when building up artwork for animation.

reverse phasing
Electronic inversion of the tonal scale in a monochrome television picture, so that white becomes black and vice versa. Useful in dealing with materials which are easier to handle in a tone directly inverse to that desired in the programme.

reverse scanning
Lateral inversion (producing a mirror image) by electronic means. Useful when, for practical reasons, a camera is compelled to view a scene through a mirror.

roller captions

A stationary camera shoots the moving area of paper between two rollers. Most commonly used for programme credits.

rushes

The first working prints taken off a negative which the film editor uses for assembling the material shot on location into coherent sequences.

safe area

Sometimes a 'safe area' is marked on studio monitors indicating cut-off (see above) so that producers can be sure essential material is reaching the audience.

screen size

Television pictures are currently viewed on a small screen. The average screen size viewed at an average distance is only equivalent to a 12 × 16 cm sheet of paper held at arm's length. Important when considering how artwork will actually appear.

scrapeback

Simple technique for making lines 'grow', without elaborate artwork. The lines are painted on cel, positioned on the animation table and filmed with the film stock travelling in reverse. Parts of the artwork are scraped off between exposures. The film is then projected forwards.

slide scanner

Comprises two 35 mm slide projectors and a simple television camera. The camera views slides from each projector in turn. As the resulting picture signals pass through the vision mixing panel (see below), cuts and dissolves between slides can be achieved.

split-screen effects

Made possible by inlay (see above). Useful when more than one item of visual information needs to be presented, without intercutting between cameras. Graphics for split-screen use have to be meticulously designed to fit previously agreed framings.

storyboard

Prepared by a designer to embody sketches outlining proposed movements and changes of scene, related commentary, music or sound effects.

strap easel

Large stand for displaying studio animations or captions which need to be larger than 12 × 9 in.

superimposition

Used to mix an image from one source with a second image from a different origin. Most commonly used for 'labelling', e.g. for identifying a studio speaker by superimposing his name. Lettering for superimposition must be prepared on a black background, and in monochrome systems should be white.

telecine equipment

Converts projected cine-film images into television pictures. For the graphic designer, opens up the possibility of working with film, particularly for animated sequences.

three-tone animation

Can be achieved in monochrome by superimposing, over an even grey background, one white on black animated caption, simultaneously routing a second white on black animated caption through inlay (see above), and changing the white information to black. Calls for perfect registration and a great deal of studio rehearsal time.

timing sheet

see dope-sheet above

tone scale

see grey scale above

type size

For domestic viewing, the minimum typographic character height will be no less than 1/25 of the total picture height. This means that on 12 × 9 in. artwork the minimum type size will be 36 point.

video-animation

A television camera is used in a similar way to a film animation camera, recording a scene into which new pictorial elements are successively introduced, producing the impression of movement. One of the advantages over film animation is the facility of instant playback via a video-tape recorder (see below).

video-tape recorder

Records sound and picture signals on magnetic tape. Expensive programmes can be justified if, by this means, they can reach a wider audience as the result of repeat broadcasts.

viewing distance

Largely determined by size of screen. Optimum has been established as twelve times the width of the screen.

vision mixing

When more than one camera is in simultaneous use, the producer can select the particular shot he requires by means of a vision mixing mechanism.

Index